CAMBRIDGE LIBRARY COLLECTION

Books of enduring scholarly value

Technology

The focus of this series is engineering, broadly construed. It covers technological innovation from a range of periods and cultures, but centres on the technological achievements of the industrial era in the West, particularly in the nineteenth century, as understood by their contemporaries. Infrastructure is one major focus, covering the building of railways and canals, bridges and tunnels, land drainage, the laying of submarine cables, and the construction of docks and lighthouses. Other key topics include developments in industrial and manufacturing fields such as mining technology, the production of iron and steel, the use of steam power, and chemical processes such as photography and textile dyes.

Memoirs and Letters of Sidney Gilchrist Thomas, Inventor

First published in 1891, this memoir describes the life of the metallurgist and inventor Sidney Gilchrist Thomas (1850–1885), best-known for discovering the method of eliminating phosphorus from pig iron which revolutionised the commercial production of steel. Professing a desire to give a 'true' account of a life in contrast to the somewhat hagiographic approach of some contemporary writers, Thomas' biographer, R. W. Burnie, sets out to construct 'a brief history of a very striking and individual character'. The details of Thomas' short life are narrated in 22 chapters, beginning with his early education, his work as a schoolmaster and police clerk whilst studying law and chemistry at night, his career, and his work-related travels, which took him everywhere from central Europe to New Zealand. The memoir also includes a postscript which reveals that Thomas left his considerable fortune to workers in steel production.

Memoirs
and Letters of
Sidney Gilchrist Thomas,
Inventor

EDITED BY R.W. BURNIE

CAMBRIDGE
UNIVERSITY PRESS

CAMBRIDGE UNIVERSITY PRESS

Cambridge, New York, Melbourne, Madrid, Cape Town, Singapore,
São Paolo, Delhi, Dubai, Tokyo, Mexico City

Published in the United States of America by Cambridge University Press, New York

www.cambridge.org
Information on this title: www.cambridge.org/9781108026918

© in this compilation Cambridge University Press 2011

This edition first published 1891
This digitally printed version 2011

ISBN 978-1-108-02691-8 Paperback

SIDNEY GILCHRIST THOMAS

Yours very truly
Sid Gilchrist Thomas

MEMOIR AND LETTERS

OF

SIDNEY GILCHRIST THOMAS

INVENTOR

EDITED BY

R. W. BURNIE

OF THE MIDDLE TEMPLE, BARRISTER-AT-LAW

'Life 's more than breath, or the quick round of blood;
'Tis a great spirit and a fiery heart.
We live in deeds, not years; in thoughts, not breaths;
In feelings, not in figures on a dial.
We should count time by heart throbs;
He most lives who thinks most, feels the noblest,
Acts the best'   FESTUS

WITH PORTRAITS

LONDON
JOHN MURRAY, ALBEMARLE STREET
1891

EDITOR'S PREFACE

IN the following pages I have sought, with what success I know not, to construct out of material sufficiently abundant, a brief history of a very striking and individual character, and of a life cut short prematurely enough, yet possessed in its own way of a singular-completeness.

It is hoped that no one who may read this little book will so misapprehend its intention as to look upon it as a glorification of personal success or money-getting achievement, after the fashion possibly of some biographies of inventors, biographies haply more grateful to the last generation than to us who stand (as it seems to some) on the threshold of a New Age. No one would more have recoiled from being ranked among the devotees of Ruskin's ' Goddess of Getting on ' than the subject of this Memoir.

Sidney Gilchrist Thomas (although placed among conditions by no means favourable for such purposes, and with working hours occupied by distasteful and monotonous business) solved a great scientific problem—the dephosphorisation of pig iron in the Bessemer and Siemens-Martin processes—and for such solution was fortunate (perhaps we should rather say foreseeing) enough to gather a pecuniary reward which, rightly or wrongly, he never regarded as his own, but rather, according to his lights, as trust-money for toilers and labourers.

Not on this account, however, is his story told here, but because it has seemed well to those who knew him, that some record should be kept of a remarkable and interesting personality, typical indeed in some ways of the very best side of our 'industrial' century, yet touched with a human sympathy which we may hope will be more general in the future than it has been in the past.

It may be observed that an endeavour has been made simply to paint a portrait, without allowing the temperament or opinions of the present writer or of anyone else to affect the rigid accuracy of the presentment.

R. W. BURNIE.

CONTENTS

CHAPTER I

EARLY DAYS

CHAPTER II

A SUMMER TOUR

CHAPTER III

A ' DOUBLE LIFE '

CHAPTER IV

THE PROBLEM OF DEPHOSPHORISATION

CHAPTER V

YEARS OF EQUIPMENT

CHAPTER VI

THE PROBLEM THEORETICALLY SOLVED—A GERMAN TOUR

CHAPTER VII

' TECHNICAL TRAVEL TALK '

CHAPTER VIII

EXPERIMENTS—A DASH INTO SWITZERLAND

CHAPTER IX

THE BASIC PROCESS PUBLICLY ANNOUNCED

CHAPTER X

THE BASIC PROCESS DESCRIBED

CHAPTER XI

TRIUMPH

CHAPTER XII

DÜSSELDORF—A GATHERING CLOUD

CHAPTER XIII

A VISIT TO THE UNITED STATES

CHAPTER XIV

HEALTH FAILS IN EARNEST

CHAPTER XV

SOUTH AFRICA

CHAPTER XVI

MAURITIUS AND INDIA

CHAPTER XVII

CEYLON, AND THE VOYAGE TO AUSTRALIA

CHAPTER XVIII

AUSTRALIA

CHAPTER XIX

HOMEWARD BOUND

CONTENTS [11]

CHAPTER XX

A SAD HOME-COMING AND A FLIGHT SOUTH

CHAPTER XXI

A WINTER IN ALGIERS

CHAPTER XXII

THE LAST DAYS IN PARIS

ILLUSTRATIONS

SIDNEY GILCHRIST THOMAS.

Photogravure by Annan & Swan, from a photograph by S.V. White, Reading.

MEMOIR AND LETTERS

OF

SIDNEY GILCHRIST THOMAS

CHAPTER I

EARLY DAYS

SIDNEY GILCHRIST THOMAS was born on April 16, 1850, at
Canonbury. His father was in the Civil Service, and a
Welshman. His mother (*née* Gilchrist) was the eldest
daughter of the Rev. James Gilchrist, the author of a
striking and individual little book, unknown to modern
readers, the 'Intellectual Patrimony.' James Gilchrist
was a Highlander, of keen literary tastes and eager after
Truth as he saw it, who drifted from Presbyterianism into
Unitarianism and thence reverted to orthodoxy, much to
his worldly detriment. One of his sons was Alexander
Gilchrist, the well-known and too early gone biographer
of Etty and of Blake. The important matter for us to
note is that Sidney Thomas was mainly of Celtic strain,
and furnished yet another example of the often unrecog-
nised addition of fame which that great race has brought
to the 'English' people.

His childhood was passed on the banks of the New
River when there was still something of a rural character

about that artificial stream. The miles of houses which now stretch over the northern slopes of the great parish of Islington away to Highgate Hill and the very gates of the Alexandra Palace were, forty years ago, still for the most part in the future.

'For the first few years of Sidney's life,' says his mother, 'he was a constant care; his brain seemed too big for his body. He learnt to read at a most unusually early age. When quite a little boy, six or seven years old, he already read much and earnestly. He would act out, in his small way, the characters of the heroes of his books—now it might be Nelson, now King Arthur, or one of the Round Table Knights. I remember, when he was seven, making for him a suit of armour, as he firmly believed it to be. Clothed in it, he would solemnly "keep vigil," pacing up and down, his sword by his side, for hours together, before making his vows to an imaginary King. One of his favourite books was a little volume I gave him on his sixth birthday — "Our Soldiers and Sailors" — short sketches of eminent men in those lines. I can see now the earnest, large-eyed child, and his delight with his presents; especially with his books. He was so rational and good a boy that his father and I thought he should by-and-bye be a clergyman. Very early in his boyhood, however, he told me with decision that that he should never be, "he was not good enough." "I will do something great, mamma, and you shall have a carriage to ride in" (I was not very strong just then), "and money to help people with."'

Sidney's mother taught both him and his elder brother (the late Dr. Llewellyn Thomas, of Weymouth Street) during their early years. When Sidney was eight he attended for a year, with Llewellyn, at the school kept in the neighbourhood by Mr. Darnell, of copybook fame.

At the end of that year Mr. Thomas removed to Grove

Lane, Camberwell, near the brow of Champion Hill, mainly that he might gain for his boys the advantage of the education given by the newly reconstructed Dulwich College, then under Dr. Carver's head-mastership. There for the next seven years Sidney remained, gradually rising from form to form in the school till the proud eminence of the 'sixth' was reached. Living at home, but attending daily at the College, the brothers enjoyed all the undoubted benefits of what is called 'home education,' together with whatever is really useful in 'public-school' life. The home in Grove Lane (well remembered by the present writer) was no ordinary educative influence. Sidney's father was no ordinary man. His talents were at once intellectual and practical, and his interest in his sons' development was ever present. Sidney was naturally precocious, and the keen hunger after knowledge (which was as much his characteristic at thirteen as afterwards at thirty) was encouraged and stimulated in every way. The boys were early admitted on equal terms to conversation both with their mother and with their father. Gossip was little favoured in the family circle. The discussion (for real discussion it would be) of literature and politics was pre-ferred to vain personal talk. Mr. Thomas himself was a Conservative in creed, his wife a Liberal by inheritance, but their sons were ever warned from accepting any opinion they had not tested for themselves, and the freest spirit of inquiry was not only welcomed but expected from them. It may truly be said that a thoroughly scientific mental attitude was thus, unconsciously to them-selves, induced in them. Omnivorous reading was the habit of the whole household.

'Sidney's mind,' says his mother, 'was stored with the kind of knowledge boys gain in a cultured home. His father habitually read aloud to the boys bits of Words-

worth, lives of great men, passages from Buffon's "Natural
History." I well remember how Sidney's cheeks glowed
at hearing read in this way the "Morte d'Arthur" of
Tennyson. He and his brother had a healthy source of
education in the visits they were accustomed to pay once
or twice a year to the country. At Christmastide and in
the early spring time, they would be received as indulged
nephews by a kind, broad-minded, busy uncle into his
Berkshire home. Here they would see the practical working of many rural industries.

'In the late summer or early autumn, they would visit
one or other of two ideal vicarage-houses. One was
Corwen, situate on the side of the lovely Berwyn Mountains, with the river Dee flowing silently and darkly on
the other side of the Holyhead high road. Here reigned
a grand old vicar, living a life of lettered dignity, and
ruling his church, his house and the parish with perfectly
absolute sway, yet with real sympathy and love. The
other vicarage was that of Llandrillo in Rhus (near
Colwyn), where the Rev. Thomas Hughes (a bachelor
nephew of him of Corwen) was vicar. This was a home
still more entirely Sidney's. From eleven to sixteen he
was a regular autumnal visitor here, and a great favourite
with the tall, hearty, breezy Mr. Hughes, the very sound
of whose laugh did one good and inspired immediate confidence. At thirteen Sidney began helping the vicar
during his visits by reading the lessons in church for
him—in the earlier days in English, afterwards, with some
training, in Welsh. These holidays, after the close work
of school, were a real blessing to him, and here he dreamed
out many an ambition for the future. Noble scenery, the
sea, books, the simple vicarage life—all these things were
a rare refreshment to the quiet, self-contained boy. I
remember a characteristic story of him at this period. A

Dean (whose cathedral I forget) was lunching with the
vicar. This dignitary put Sidney through an examination
in Latin. The boy came so well out of the ordeal that
the Dean "tipped" him three half-sovereigns and retained
him as guide over the Great Orme's Head. A happy hour
ensued; Sidney in the heat of some discussion flinging
off his jacket and carrying it under his arm. The half-
sovereigns had been put into the jacket pocket, and not
unnaturally, upon return to the vicarage, they were gone.
Not unnaturally either, the vicar was vexed; but Sidney's
only answer was: "Never mind, godfather, most likely
someone has it who wants it more than I." "What can
you do with such a boy?" wrote the vicar to me.

' Never did he as a lad care for money in the way boys
often do. Once, some money having been given him, he came
to his father and offered him five shillings for a little worn-
out American clock. His father told him the clock was
not worth the money and that he might have it for
nothing. Sidney, however, said that he wanted to take
the clock to pieces, and must therefore pay for it. Take
it to pieces he did, and, not being a watchmaker, was
naturally unable to put it together again. He remained,
nevertheless, perfectly content with his bargain.'

Constant discussion of political questions, coupled with
unceasing insistence by his elders that he should render
a reason for the faith that was in him, made Sidney a
militant Radical at an age when it may be supposed that
most boys are chiefly interested in cricket stumps and
footballs, not to say in tops and marbles.

From the beginning he followed the course of the
American Civil War with the eagerness and comprehen-
sion of an intelligent man. Alone in the family circle he
would do battle for the North, and upon fitting occasions
(for he must not be supposed to have been in any sense

that most dreadful of social plagues, an 'infant phenomenon') would argue on State Rights and what not with a
knowledge and an accuracy which would have done credit
to a disputant thrice his years.

The last sentence leads us specially to emphasise what
is necessary to be remembered in connection with what
has gone before, that amid all this precocity no element of
priggism was allowed to intrude. The slightest flavour
of this detestable spirit would have been instantly detected
and unsparingly ridiculed. Sidney was before all things
trained to be a boy while boyhood lasted. Nor was it
desired to cultivate mental at the expense of physical
faculties. Open-air pursuits and recreations were encouraged in every way. Each Sunday afternoon Mr. Thomas
would take his boys long country walks, by no means
restricting himself to the high roads, but striking ' across
country' whenever opportunity offered. On these expeditions, and indeed whenever they found themselves in
fields or roads, the lads were taught to use their eyes
to good purpose. Natural history was a passion with
Llewellyn Thomas, and Sidney also cultivated it in a
minor degree. Thirty years ago Camberwell (or the
up-lying portion of it at least) was still on the edge of the
country, and abundant opportunity was to be found for
entomological collecting—even for birds'-nesting on a
somewhat extensive scale, and with a more or less scientific object.

A well-thumbed copy of a little book by Mr. Atkinson
on ' British Birds' Eggs and Nests' was a classic in the
home. Llewellyn and Sidney were joint possessors of a
regular aviary, with a constant population of some dozen
birds of different species, an intense source of delight to
both boys. For some time an effort was made to keep
a kite in the garden; but the bird developed such an

unpleasant habit of attacking innocent visitors, that
ultimately it was deemed necessary by the domestic
authorities to cause its presentation to the Zoological
Gardens. These tastes were adopted by Sidney (so far
as they were adopted at all) in emulation of his elder
brother. From very early days his own individual predi-
lections took a different direction. Mechanics and engineering had an ·irresistible fascination for him from the
time when (κτῆμα ἐs ἀεί as it seemed indeed) he became
possessor of his first box of tools and fashioned his first
toy ship. A little later, in 1862, during the formation by
the then youthful Metropolitan Board of Works of the
New Main Drainage System, he would stand for hours on
a half-holiday entranced in contemplation of the building
of the great sewers. He soon decided that his avocation
in life was to be that of a mechanical engineer. A year
or two afterwards the fairy'and of chemistry opened before
him, and he resolved upon becoming an analytical
chemist. Little did any then foresee the devious paths by
which he was to be led back to his first mistress, Science.

Art, however, had also its influence on the boy. The
Dulwich Gallery was a favourite resort during recreation
hours at the neighbouring college. Every picture in
the collection was known by heart, so to speak, to Sidney,
and its history and every fact connected with it. Music,
too (although in after life he always disclaimed special
liking for modern developments on Wagnerian lines, or,
indeed, any special taste for it), had always in truth a great
attraction for Thomas. At fourteen the wonderful singing
and playing of Miss Havergal (a lady whose religious
verse made her quite famous at one time in certain circles)
produced a strong impression on him, 'as well,' says his
mother, 'as the deep spiritual individuality of the sweet
singer herself.'

Amid all these influences and dreams, the steady,
regular school work and life at Dulwich maintained an
admirable balance of compensation. Sidney was boyish
enough in all conscience when joining on a summer afternoon or evening in a hare and hounds paper chase round
the borders of South London. Over the whole scheme of
education presided a steady inculcation of industry and
energy in all things, whether work or play, very delightful to witness. To use an expressive Americanism, the
household at Grove Lane was a 'live' household, with no
particle of sullen sloth about it. Self-reliance was one of
the earliest lessons taught the boys, and at twelve years
old or less, they were expected to be able, unassisted, to
escort a less experienced country cousin to a day's sight-
seeing in town, or with equal facility to join him in a day's
birds'-nesting in the country.

'From the time Sidney entered Dulwich,' says his
mother, 'his progress was steady. He was always obedient,
always industrious, yet seeming to lead an inner life of his
own. I remember that at fourteen he had a vehement
struggle with another boy for the top of the fifth form.
Especially was their competition keen for that form's prize
for Latin Verse and Prose Composition. This prize Sidney
gained. Comparing notes afterwards, however, with his
friendly rival, he came to the conclusion that it had not
been rightfully adjudged to him. No sooner was he convinced of this than he sought an interview with the
Master, and endeavoured to convince that authority that
the decision was wrong. The Master was both amused
and aggravated, and told Sidney that he had better be
content with what praise and success were given him
in this hard world. Nevertheless, Sidney remained
thoroughly dissatisfied with his victory, taking no pleasure
in his prize.

' His protecting love for his little sister Lilian, eight years younger than himself, was born with her birth and grew with her growth. When she was a week old he would ask the nurse to be allowed to take her in his arms, and upon the good woman's consent, would sit holding the baby-sister for half an hour at a time, never moving, but silently looking at her. As Lilian grew older, Sidney became her companion and friend, teaching her, telling her fairy tales ; upon returning from an absence always bringing her some little memento of the spot visited, or some odd quaint tale of adventure.'

Equal with Sidney's love for his sister was his devotion to his mother.

' One of the strongest ties of his life,' says the latter, ' was his devoted affection to me. When he was fourteen he had a serious illness, inflammation of the lungs and brain, brought on (so the doctors said) by overwork, and by carelessly getting wet in walking across the fields to school. (At that time there were fields between Camberwell and Dulwich, and not streets of speculative builders' masterpieces.) Through this terrible illness I nursed him. He and I were shut up together for three anxious months, and our mutual affection and devotion were, if possible, strengthened. During his convalescence from this malady he would sometimes give me a glimpse of his inner thoughts. Through science (always through science) he was to do some great thing, and Lilian and I were to help him to dispense among the unfortunate and the neglected the money he was sure to make.

' When Sidney had attained his sixteenth year, Dr. Carver, the head-master of Dulwich, wrote to my husband requesting an interview. At the meeting which thereupon ensued, Dr. Carver said that he was most anxious that Sidney, who he thought would do honour to the school, should remain some time longer at Dulwich, and should ultimately go in for a scholarship at either Oxford or Cambridge. To such a scheme my husband had no objection; on the contrary, he was eager for its execution. "Sidney," he used to say, "will in the end become a man of science; but he will be a credit to whichever university he may join. One thing, however, is certain : money will never be an object to him; indeed, he will never be able to take care of it." This last prediction the future was fated signally to falsify.

'However, an insurmountable obstacle arose to all these plans. Sidney, in his own quiet, respectful way, told both the head-master and his father that he would rather matriculate at London University and study medicine in the capital.

'Dr. Carver, his father, I myself, were all much disappointed; but the boy had his way. In the summer holidays of 1866 he left Dulwich. In that summer, too, he accompanied his father upon a long tour in South Wales.

'That trip strengthened the boy's affection for his father, and more than ever convinced the latter that he had a rare nature to deal with. Upon their return my husband said: "Sidney can pursue his own course; we can absolutely trust him."

'My boy, on his side, at once began studying for the London Matriculation. His father offered him a coach. "No; please, father," said he, "a fellow knows nothing really well which he does not gain for himself." So passed the weeks, Sidney working up his subjects himself, and also devoting his time to teaching Lilian and his younger brother Arthur. He began Latin with them; made geography lessons easy to them by telling them tales of strange countries. Always the instruction was wound up by some wonderful story invented for the occasion.'

In such fashion were the irresistible forces of heredity and of education combining to mould a bright, alert, questioning, indefatigable, strenuous, and withal practical spirit. A sudden family crisis was to test that spirit earlier than had seemed likely. Dreams of matriculation at London, of study of medicine, of ultimate pursuit, mayhap, of analytical chemistry or mechanical engineering, were to disappear.

In February, 1867, Mr. Thomas died suddenly of apoplexy, and the household was left without a head. The loss of income was naturally serious. Llewellyn, the eldest son, had already entered upon his career (a career destined to prove brilliant enough, although cut short too early), and for a twelvemonth past had been attending at St. Thomas's Hospital. Sidney's resolves had better be told in his mother's words :

' Sidney sat down by his father's bedside a boy ; from his grave he passed out a man, and thenceforward took upon himself, as far as he could, the burden of my grief. When we were alone, he told me quietly that he should not matriculate, that he should write to the vicar of Llandrillo, and endeavour to obtain a Civil Service nomination ; that he would take anything that first offered. I prayed him to carry out his plans. I said we would all live quietly together, and that we should have income enough. "Mother," his answer was, "you will want all you have to educate the little ones."

' No prayers, no argument could move him, and so this boy (not yet seventeen) launched himself on a man's career. He wrote to his Llandrillo cousin and godfather, and had a speedy promise of his nomination. In the meantime, Sidney devoted himself to urgent affairs. His

father had been executor of the vicar of Corwen, who had
died the preceding summer. He carried on this executor-
ship and helped me with my own.

'Shortly after my husband's death we moved to
Camberwell Grove. We had not long settled there when
Sidney told me quietly that he had taken a classical
mastership at an Essex school, meaning to hold it until
the promised Civil Service appointment came. He explained that there was nothing now for him to help me
in. "You know, mother, I cannot be idle."

'No remonstrances availed. He went to the Essex
school—it was at Braintree—and found his class to consist
of young fellows bigger and older, for the most part, than
himself. These lads were at first much inclined to re-
bellion : but Sidney persevered, prevailed, and in the end
reduced them to willing obedience. The head-master was
most anxious to secure his classical assistant permanently,
and offered him increased salary and ultimately partnership
if he would remain.'

However, the particular drudgery of teaching was
always abhorrent to Sidney, tolerant as he was of drudgery
when needful, and he was by no means ill-pleased when
the looked-for nomination came. It was to a clerkship in
the Metropolitan Police Courts.

Attached to each Metropolitan Police Court are a
'senior' and a 'junior' clerk, members of the Civil Service.
The junior's salary begins at 90l. a year, with an annual
increment until 200l. a year is reached ; the senior receives
500l. per annum. The seniors are recruited from the
ranks of the juniors ; but in so small a department promotion is necessarily slow, and the discoverer of the
Thomas-Gilchrist process never attained it in his twelve
years' service. The duties of the clerks are to conduct all
the business of the office as distinct from the Court, to

receive and account for all the moneys paid in for process,
fines, &c., and in court to take notes and depositions.
The examination of witnesses, in the great majority of
cases where no advocate appears, is by most magistrates
left much to the clerk. To anyone with the slightest
knowledge of the volume of business constantly transacted
before these tribunals, it will be obvious that the official
hours from ten to five must be pretty fully occupied. At
the busier courts, indeed, the clerks are often detained an
hour or so later, although the magistrate himself, of course,
adjourns at the statutory time. This is mentioned for a
reason which will presently appear. Thomas, having obtained his 'nomination,' had little difficulty in success
in the examination, with some hundreds of marks to
spare. A year or two later, equal success in his
examination would have given him to a great extent his
choice of departments. At this time, however, it was not
so. In the latter part of 1867 he entered upon his duties
at the Marlborough Street Police Court. Mr. Knox was
the senior magistrate here at the time. The work was
quite novel to Sidney; but, although he never liked it
(indeed, disliked it cordially), he buckled to it with
characteristic energy. At any rate, it was better than
teaching. It is not too much to say that, in the midst of
all the other more congenial pursuits of which we shall
presently speak, he found time to thoroughly master not
only the practice and procedure, and the various statutes
with which he was more immediately concerned, but, indeed, to make himself an accomplished criminal lawyer.
In the earlier days at Marlborough Street the atmosphere
was, doubtless, strange enough to him, and the writer can
well remember his telling with much gusto how he tried
to convince Mr. Knox that he should not convict a
man who, when starving, had appropriated another's loaf,

because even so conservative a thinker as Paley had maintained that such a taking was not theft. The worthy
magistrate was puzzled for the moment by this citation of
an authority so little quoted in law courts, but presently
bethought him that in truth the plea of necessity could
hardly arise, since the merciful legislation of this happy
country had provided for the destitute the pleasant asylum
of the casual ward.

Marlborough Street is probably the police court where
the work is lightest, and it is situate in a locality which
is accessible and agreeable to the average middle-class
man; consequently the ordinary police-court clerk seeks
rather eagerly after appointment to it. Sidney, however,
was neither an average middle-class man nor an ordinary
police-court clerk.

In 1868 the East End had not yet been discovered
by Mr. Walter Besant. Nobody knew of the delightful
pastime styled ' slumming; ' nobody dreamt of Palaces of
Delight, or produced glorified technical schools. Thomas
was nevertheless smitten with a genuine desire (since
police-court drudgery seemed to be his portion) to pursue
his vocation rather in the East than in the West, and to
see for himself something of the great depths below our
civilisation. For probably the first time in the history of
this branch of the Civil Service, he sought an exchange
with a colleague at the ' Thames ' Court in Arbour Square,
and naturally met with no difficulty or obstacle in the
achievement of his wish. He thus quitted the West End
Court after about a year spent there, and for the remainder
of his time in the profession was attached to the Stepney
tribunal. Among the magistrates here were Mr. Paget,
Mr. De Rutzen, Mr. Lushington, and for a short time
before Sidney's resignation, Mr. Saunders. Thomas continued to live at home. His mother, as we have seen, had

removed from Grove Lane to the neighbouring 'Camber-well Grove.' Naturally, and gradually, while still little more than a boy, he assumed unconsciously the position of head of the family; for his elder brother was by this time out in the world on his own account, and no longer a constant member of the home circle. He would usually walk the long distance from Camberwell to Stepney at a swinging pace, always arriving at the Court at ten sharp; often, indeed, he would walk back. At Thames he had a senior colleague, a Mr. Poyer, since deceased. With this gentleman Sidney was enabled, after some years, to make an arrangement which left him two days a week free, and this gave him precious time which was devoted to the real mistress of his heart, Science, and to study and researches by means of which he, in the end, perfected that which was to prove his life-work.

Before we speak of this pact, which had so much influence on the future, let us here introduce a description of Thomas as he appeared at this time to a cousin and intimate friend, who took a holiday tour with him in the summer of 1869.

CHAPTER II

A SUMMER TOUR

' IT was in the gorgeous July of 1869 that Sidney Thomas
and I, he then being aged nineteen and I a year or two
younger, visited the Continent for the first time. Such a
visit at such an age is an experience never in any case to
be forgotten ; but in this instance my cousin's striking
personality must, anyhow, have indelibly impressed upon
one's mind all the main incidents of a month's travel with
him. During our walks along the straight white Norman
roads we discoursed " of all things, and some others," with
that wonderful self-confidence—alas! also with that wonderful energy and new delight—characteristic of the dawn-
ing days of manhood, when life is like a romance " of cloak
and sword," and not the dreary, grimy, realistic narrative
which it too often afterwards becomes.

' We were, I think, both possessed of that keen pleasure
in argument, for the sake of argument, which older out-
siders sometimes find so distasteful to them in smart lads
in their teens, and we naturally always took opposite
views of every conceivable topic, from the mysteries of
theology down to the topography of the Lower Seine. The
summer air would be heavy with the clang of debate as
we trudged along. Yet we had, I think, both of us, a
wonderfully happy time of it, and as light hearts as any
pair of youngsters in all fair France. Light hearts have
a proverbial accompaniment, which in our case was not

lacking either—to wit, light purses; but need for economy, provided it be not too pronounced, only adds to the enjoyment of a pleasure-trip at twenty.

' Of the well-remembered little incidents of that trip, so far as they illustrate either Sidney Thomas' character as it appeared to me, or the experiences which were going to form it, I will say something presently; but I want, if I can, in the first place, to give some idea of that unique personality of his at which I have already hinted. Such as he was then, such he remained, in my eyes at least, almost to the end. No one with the slightest faculty of observation could ever have come into the most momentary contact with him and have failed to recognise a mind of exceptional power. He had the spare frame of a man eager, not merely for intellectual research, but for intellectual conflict and conquest, of a man perhaps somewhat too disdainful of the things of the flesh. His face was a little " sicklied o'er with the pale cast of thought " and his hair a little long and unkempt (of a surety from no conscious affectation, nor indeed had " æstheticism " begun in 1869); yet I think most women would have found his clear-cut features and speaking eyes, wonderfully variable in colour and expression, handsome. He spoke in a clear, pleasant voice, which in moments of excitement became metallic. His reading was wonderful for a youth of his age—fiction, history, travel, theology, on all these subjects he seemed equally at home. Perhaps poetry had been a little neglected. In the semi-humorous, self-depreciatory way which became him well, he used to say that he had no care for verse, and that in the coming time everything worth reading would be written in prose; but I never believed either assertion. Social subjects had a wonderful fascination for him, and although his mind was too independent to accept blindfold any of the provisional theories of

C

the human future which had come in his way, and he was
"nullius addictus jurare in verba magistri," yet I do not
think, looking back across the expanse of twenty years,
that it would be saying too much to describe him as almost
persuaded to be a Socialist. I know that in those days he
was far more advanced than I, who had but faint glimmer-
ings of social problems; although politically I was radical
enough. Of science he seldom spoke to me, knowing how
feeble my interest in and scant my knowledge of those
departments of it, at least, which specially attracted him.

'Under the stimulus of what to us were novel experiences in wayside Norman inns or on the asphalte of peer-
less Paris, sides of Thomas's character became apparent
which were not so well seen in his workaday life, when
he was subjecting himself to that double strain of distasteful exertion conscientiously performed in the fetid
atmosphere of a London police court and congenial study
unfortunately pursued in hours which immutable hygienic
laws have decided should be devoted to leisure. Most of
us, who belong to the non-productive classes at least, know
nowadays something of the mental exaltation produced
by realising for the first time with our own eyes the
existence of a civilisation different from our own, even if
it be only the civilisation of a country so like ours as is
France. The very names on the shop-fronts, the very
jabber of the children in the streets, the very knowledge
that we are strangers and sojourners,—all those things
cause a delight never afterwards to be reproduced. For
myself, I shall never forget our landing at Havre one
afternoon in early July. We had come by the long route
from London Bridge, and I think we had both of us
suffered a good deal in the Channel. All the morning we
had lain tossing outside the harbour waiting for the tide.
Such troubles were soon forgotten as, in our phrase-book

French, we asked our way, knapsacks on back, to the Caudebec road ; for we were to walk up the Seine valley, Paris-ward.

'With what zest we ate our rolls and drank our *café au lait* in the morning and felt that we were indeed "on the Continent!" We did a good trudge that day, I remember. Thomas resolutely refused to eat any *déjeuner*, a resolution which he adhered to pretty steadfastly throughout our travels, maintaining that our rolls in the morning, with our dinner in the evening, sufficed for all our needs. This was a doctrine which I as steadfastly opposed, insisting on the midday repast as a necessity. Hence arguments which speedily led us far afield over the whole domain of what we knew of physiology, and from physiology the way was easy to dispute concerning most things in heaven and earth. The echo of our words comes back to me now, with the background of the straight white roads, the hedgeless fields, the kilomètre-stones, and the iron guide-posts. I did not know of the purposes which were even then doubtless dimly shaping themselves in Sidney's mind, and leading him to a settled scheme of minute economy in his expenditure upon himself, so that, when the hour struck, he might not fail in his projects for want of the sinews of war.

'There was, I have always thought, however, joined with this intelligible motive to abstinence, a half-conscious leaning to asceticism in Sidney's nature which impelled him to unnecessary and even injurious self-denial. I much fear that the seeds of premature decay were implanted in his naturally vigorous frame by the habit which he acquired in these adolescent years, when abundance of food is of prime need, of systematic under-eating—a habit, the evil results of which were assisted, as has already been hinted, by systematic over-work. But these things were absolutely hidden from us by futurity's curtain, nor did any anticipation of evil to come spoil our summer days.

' Paris in the midsummer of 1869 seemed to our inexperienced eyes the City of Pleasure in very truth, and doubtless we missed the lessons we might have learnt in the streets of the City of Light. In little more than another twelvemonth, the frequenters of the boulevard, with their English-made clothes and their twisted moustaches, would for the most part have fled elsewhither; but the real children of Paris, noblest populace perhaps of the world, would be enduring with fortitude, never before shown by such a mass of human beings, all the horrors of the long siege. In some twenty months' time, those same children of Paris would kindle a flame which should terrify respectable persons everywhere, and be as a beacon to lighten the steps of revolutionists for many a day.

' Although we did not dip much below the surface, we crowded a great deal of sight-seeing into our eleven days in the capital. Sidney was, as ever, insatiable after new things, and, although never tired of satirising himself for the foible, yet was seemingly bent on emulating the typical Yankee anxiety to fill the day with achievement.

' On our return walk from Paris to Dieppe I remember a country gendarme stopped us once and demanded our passports. We said that we were English and needed none; but "Je crois que vous êtes des Prussiens," rejoined the moustached and swaggering Dogberry. However, after some consideration he allowed us to go our ways, yet still with scowling mien walked his horse after us for a kilomètre or two, until, I presume, we passed out of his jurisdiction.

' We disliked this dogging of our footsteps very much, and at Sidney's suggestion we started the " Marseillaise," feeling all the time that we were very desperate ruffians indeed ; but as we could neither of us sing a note, and as we knew nothing of the tune, and but little of the words of the then forbidden song, I really do not think that our persecutor realised our audacity. Another time we walked some miles with an ex-convict from Toulon, in whom Thomas took much interest, but from whom we gathered little save a general impression that our interlocutor was a well-meaning, stupid fellow, somewhat dazed with the injustice of the world.

'We were absent a month, and out of the ten pounds apiece we had started with I brought back some sixteen francs, but Sidney double or treble that amount. Had it not been for his example, I should never have done things so cheaply. I insist on these details because Sidney's severe and rigid, perhaps too severe and rigid, economy throws much light on some main features of his character. We may hope that in the better society which the future, as some of us hold, has in store for us, thrift may cease to be deemed a virtue ; since, where each one renders according to his capacity and receives according to his needs, there will be no fear of ever wanting. But under the present false social conditions, and in the horrible world in which we live to-day, there is, it seems to me, revolt as we may from asceticism, no undeserved credit due to him who, for a worthy and unselfish purpose, not only " shuns delights and lives laborious days," but even by abstinence hoards out of scanty means the wherewithal to battle hereafter. Thomas was no miser, and no man more generous to others ever lived. He only pinched himself.

'He had, as it seemed, an inborn financial genius. Perhaps this was merely a manifestation of his keen sense of things as they really were. His imagination was powerful enough in some directions ; but it was always his servant and never his master, and his outlook on the world was quite unobscured by mists of fantasy or passion. Yet none was bolder in speculation, and in many matters he was an idealist. I will not say that he had quite " swallowed all formulas "—few of us, strive as we may, succeed altogether in that; but he had proved most things, and he held fast those which seemed to him good.

'Looking back on these somewhat rough notes— wherein I have endeavoured, perhaps not too successfully, to paint my cousin's portrait in rather "impressionist" fashion—it seems to me that I have given, it may be, too harsh and stern a rendering of one of the most genial men I ever knew. Stern and even harsh he could be upon occasions, although never for long; but habitually he was the most cheerful, the most fascinating, even the most humorous and lightsome of mortals.'

CHAPTER III

A 'DOUBLE LIFE'

IN the foregoing chapter we have Sidney Thomas as he
appeared to an intimate friend when on holiday-making
bent. At home he had become practically, as we have
said, the head of the family, his elder brother being out
in the world. After the removal from Grove Lane to
Camberwell Grove, there began, says Sidney's mother, ' a
new domestic life, of which Sidney was the centre.' His
official work at this time (1867–1871) was hard enough,
as indeed it always was, and the two free days a week—
to be by him devoted to still harder scientific work—which
he subsequently acquired by arrangement with Mr. Poyer,
were as yet in the dim and distant future. Hard as might
be his police-court labours, unattractive to him as they
often were, he threw his whole heart and soul into their
discharge. Always an early riser, he had mastered the
morning paper, eaten his breakfast, done miscellaneous
work, and walked, as his usual manner was, the long miles
from Camberwell to Stepney easily by ten o'clock.

There, day after day, he would arrive with ever-fresh
energy, always buoyant with a vitality which, so long as
he remained at the court, was to the very end entirely
devoted to his official duties. Mr. Lushington, under
whom he served for ten years, brings out well this buoyant
energy, which was one of Sidney's most marked charac-

teristics, in the following letter addressed to Sidney's sister,
now Mrs. Percy Thompson:

'Thames Police Court: January 1890.

'Dear Miss Thomas,—Your brother, Sidney Gilchrist
Thomas, appears to have been transferred as second clerk
to this Court some time in the summer of 1868. I have
been unable to find any letter announcing the exact date
of his appointment; but his handwriting begins then to
appear in the Court Letter Book, and this would accord
very well with my own impression that he had been here
from eighteen months to two years when I came to the
Court in December 1869. He left it in 1879, so that I
had the pleasure of his help for nearly ten years, and
enjoyed the fullest opportunities of appreciating his value
in our business relations, as well as of gaining an insight
into his character. During most of those years, the
pressure of work at the Thames Court upon the magistrates, and the clerks also, was perhaps harder and more
unremitting than at any other Court in London.

'Your brother was as indefatigable, as clear-headed, as
patient in dealing with stupid or ignorant witnesses, as
accurate and concise in putting the evidence into the form
of a deposition, as any clerk could possibly be; and he
was bright and elastic from the beginning of a long day
to the end, and from one long day to another, with work
so heavy as to require its being got through with all the
rapidity that was compatible with efficient performance.
It was a constant help and a constant satisfaction to me
to see his part performed, not only with the exact
mechanism of a trained intellect, but with the thorough-
going industry of a conscientious and passionate lover of
strict justice.

'I instinctively felt that he formed his opinion inde-

pendently of mine, and that he was the most competent
and unbiassed, and in some ways the severest, critic of the
style in which my own duty was performed. Wherever
a touch of out-of-the-way medical or scientific jurispru-
dence came into the details of a case, I was always particularly struck with his quick appreciation of the points
in the evidence of any expert witness. I understood that
he was fond of practical chemistry; but it was not until
after the publication of his great discovery that I became
aware of his possessing a genius in that line that would lift
him at once into the first ranks of scientific reputation.

'I was most sorry when his new career removed him
from the staff of the Court, though delighted with the
extraordinary success he had achieved; and I am sure
that every official of the Thames, from the highest to the
lowest, was equally fond of him while there, equally proud
of him when he went from us, and equally grieved at his
early end. His career was an instance of the precept of
the Preacher: " Whatsoever thy hand findeth to do, do it
with thy might."

'Believe me, very truly yours,
'F. LUSHINGTON.'

Most Civil servants, after the hard collar-work, extending over seven or eight hours, which is here described,
would have thought their evenings at least sacred to re-
laxation; but Thomas was made of different stuff. His
evenings, as soon as his simple dinner was disposed of,
were always spent in work of some kind, and very soon
came to be specially set apart for chemical studies and
experiments. He early began to lead the double life—a
very virtuous 'double life'—which was to be his for a
decade at least. In one aspect and to one set of acquaintances he was a model and exceptionally intelligent police-

court clerk; in another aspect and to another class of
friends he was a promising young scientist. Into his
leisure hours he crammed work which would have been
more than sufficient for all the energies of most men.
We will return to his chemical pursuits presently.
Let us note here that, beyond all this, he had burdened
himself with the management of the financial affairs, not
only of his mother, but also of several other female relatives. The keen, practical, business-like side of him,
which was as markedly characteristic as his idealism,
delighted in threading the intricacies of the Stock Exchange, and he was a thoroughly trustworthy guide to
'investments'—never really rash, although sometimes
seeming so.

Yet he never seemed too busy for such a long talk
with a congenial friend as his soul loved, and in some
mysterious way he contrived to read more general literature of all kinds than many professed literary men.

In the summer of the fearful and memorable year
1871, Mrs. Thomas let her house in Camberwell Grove for
some months, and went abroad to Germany, Switzerland,
and Italy with her two younger children and Miss Burton,
a cousin, returning early in 1872. Sidney accompanied
them, as far as his annual holiday would stretch, and then
returned to harness.

During the absence of his family Thomas lived a
somewhat solitary life in London, residing at first in a
boarding-house in a City square, and afterwards in lodgings
in Brooke Street, Holborn.

The following extracts from letters belong to this period:

To his Mother

'1871.

'Dearest Mother,—I have just contrived to squeeze out
a moment or two to write to Lil. Square as I anticipated a

failure; but I am of course in a fix, as I can't get a day to
look about. I have seen three rooms close to Chancery
Lane, very small, dingy, only 15*l.* per annum; of course
empty. They are not empty till end of month. I calculate attendance about 4*l.* a year; light, fuel, and furniture
6*l.*, and glorious independence. No more boarding-houses
for me. However, it is uncertain whether the rooms are
not already let. London viler than ever. How I envy
you in your luxurious retreat, far removed from the toils
and cares of your deserted sons.'

To his Sister

'1871.

'Dearest Wee Maid,—How *dare* you go and spend your
Xmas away from your devoted boy, and leave his Xmas
pudding to the chances of promiscuous charity and his
own culinary skill? The truth is, I am conscious of deserv-
ing a scolding for not having rushed, with eager pen, at
once to respond, as best I might, to your two delicious
little epistles, and so hope to avoid the merited reproof by
exposing my own grievances. It is needless to remark
that I only recovered from the impression that I was the
fortunate recipient of one of the world-famed missives of
Sévigné, whose epistolary style has been chastened and
adorned during her residence with the shades by the
instruction and examples of a Lamartine, a Rochefoucauld,
and a Dumas,—I only awoke from this delusion, I say, when
I recognised the well-beloved signature of my honoured
sister. But really I was much pleased, both with your
style and expression, while your communication in the
vulgar tongue was equally acceptable and less straining to
one's intellectual department. Now I really don't know
if I am *en règle* in wishing you a Merry Xmas, which I
had intended to do; for I suppose you will spend it in a

picnic on Arno's banks, with umbrellas to keep the sun off, and an airy repast of strawberries and cream or grapes and ices, or in some other festive mode befitting the " sunny South ; " whereas we all know that roast beef and its concomitant plum indigestion, with snow on the ground, is absolutely essential to a Merrie Xmas.

'And so I, with a prospective possession of these latter blessings, look down with pity on you benighted foreigners. By-the-bye, talking of matters culinary, you have no conception what a professor in matters gastronomic this fraternal genius of yours is becoming. To see him boil a potato, roast a haunch of venison (N.B.—a frequent dish), and finally prepare his great and world-renowned dish of *omni cum omnibus bene extrare, mixta cum quibus domolii,*—oh, that indeed is a sight calculated to rejoice the spirits of a Soyer or a Francatelli ! And then to see his tranquil happiness and serene beatitude when, relieved from his pleasant toils, with heels gracefully reclining on far-upsoaring mantelshelf, and with easiest of chairs backtilted to the uttermost verge of unstable equilibrium, he rests exposed to the rays of a glowing fire, with pleasant novel and not unpleasant dreams ! Now, after this fascinating picture of life *en garçon*, don't you feel tempted to join in an alliance with this fond youth and leave the rest of the family out in the cold of the blue skies of Italy ? Post of housekeeper still open ; no one over twenty-three need apply. The midnight bell is striking, so, darling, once more a Merry Xmas and Happy New Year.—Ever yours,

'SIDNEY GILCHRIST T.'

Some readers may be astonished at some passages in the above letters. Chambers in Chancery Lane at 15*l.* a year, with attendance calculated at 4*l.* per annum and ' light, fuel, and furniture ' at 6*l.*, may seem a vain dream

of economy.   But it really was upon such bases that
Thomas arranged his existence.   His thriftiness was,
however, as his cousin explains above in the account of
the French tour of 1869, confined to his own personal expenditure, and was doubtlessly largely dictated by the
necessity of accumulating out of a small enough income
the nest-egg which would be needed for those ultimate
purposes which were shaping themselves more and more
clearly in his mind.   Under our present social system, if
a man be born in the purple, he is not likely to
revolutionise metallurgy by his discoveries; if he be not
so born, and yet have such an aim, he must not only work
night and day, but also pinch himself for years to obtain
Capital.

CHAPTER IV

THE PROBLEM OF DEPHOSPHORISATION

ALL this time Thomas's purposes were ripening. We have already told how in the very early days of 1868 he had already begun experimenting and studying at home in the evenings. In 1870 he attended a course of lectures at the Birkbeck Institution delivered by Mr. George Chaloner, who then held, as he still holds, the teachership of Chemistry at that admirable school. Sidney had from the first given himself to the examination of the unsolved problems of chemistry;[1] but it was at these lectures in all probability that he received the final impetus which started him in pursuit of a solution of the particular problem destined to be indissolubly associated with his name. Mr. Chaloner took occasion to say that ' the man who eliminated phosphorus by means of the Bessemer converter would make his fortune.' There can be no question that this expression sank deeply into Thomas's mind, and about this time he frequently quoted it. It has indeed been said ('Iron,' No. 630, p. 111) that ' the

---

[1] Although dephosphorisation of iron pig was the question to which Thomas ultimately devoted himself, yet he always kept in his mind other problems which perhaps, had he lived, he would have elucidated as triumphantly. Mr. Chaloner is wont now to tell his pupils how Thomas would repeatedly insist to him on the hydrogen, oxygen, and nitrogen present in air and water and to be had for nothing, and the little use made of them. ' Impossible as with present lights it may seem,' he would say, ' why should not ammonia be extracted from the air ? '

commercial idea here expressed was quite as much in his
thoughts as the scientific nature of the problem. In early
conversation on the subject he frequently used to point out
the product of a royalty of sixpence a ton on 3,000,000 tons
annually of Cleveland pig.' No doubt that Sidney looked
forward to the realisation of riches, should he discover
the secret of the dephosphorisation of iron in the converter. His mother has told above of his early dreams of
fortune and his visions of good purposes to which that fortune should be applied. Yet we may take leave to doubt
whether this supplies any support to the threadbare
theory that great inventions are only to be encouraged by
monetary rewards. The bent of Thomas's mind would, in
a society where money did not exist, have carried him
quite as irresistibly towards discovery—perhaps even
towards this particular discovery; the stimulus of fame,
nay, the intellectual pleasure in doing good work, would
have been quite as effectual as the desire of riches even for
others.

In any case the solution of the dephosphorisation problem
became from this time forth his chief thought and object.
We may explain here in what that problem consisted.

Up to 1855 the process of making steel from iron had
not varied for a hundred years. In the middle of the last
century a certain Cort had invented a new process, which
in its time undoubtedly marked a new departure in the
world's history. Until Cort's discovery, the finest steel
used in this country was made by the Hindoos, and is said
to have been quoted at the fantastic and prohibitive price
of 10,000l. a ton. Cort produced equally good steel at
prices ranging from 50l. to 100l. a ton. Still, even at such
prices as these what has been called the 'Steel Age' could
not be said to have begun.

That age began when Henry Bessemer, between 1856

and 1859, worked out an entirely new method of steel
manufacture, a method destined to revolutionise this most
important branch of metallurgy. By this process pig-iron
is transformed into steel by being ' blown ' in a ' converter.'
On May 24, 1859, Bessemer thus described his process to
the Institution of Civil Engineers:

'The converting vessel is mounted on an axis, at or
near the centre of gravity. It is constructed of boiler
plates, and is lined either with firebrick, road drift, or
" ganister "—a local name in Sheffield for a peculiar kind of
powdered stone, which resists the heat better than any
other material yet tried, and has also the advantage of
cheapness. The vessel, having been heated, is brought into
the requisite position to receive its charge of melted metal,
without either of the "tuyeres," or air-holes, being below
the surface. No action can therefore take place until the
vessel is turned up, so that the blast can enter through the
tuyeres. The process is thus in an instant brought into full
activity, and small, though powerful, jets of air spring
upward though the fluid mass. The air, expanding in
volume, divides itself into globules, or bursts violently
upwards, carrying with it some hundredweight of fluid
metal, which again falls into the boiling mass below. Every
part of the apparatus trembles under the violent agitation
thus produced; a roaring flame rushes from the mouth of
the vessel, and, as the process advances, it changes its
violet colour to orange, and finally to a voluminous pure
white flame. The sparks, which at first were large, like
those of ordinary foundry iron, change into small hissing
points, and these gradually give way to soft floating specks
of bluish light, as the state of malleable iron is approached.
There is no eruption of cinder as in the early experiments,
although it is formed during the process; the improved
shape of the converter causes it to be retained, and it not

only acts beneficially on the metal, but it helps to confine the heat, which during the process has rapidly risen from the comparatively low temperature of melted pig-iron to one vastly greater than the highest known welding heats, by which malleable iron only becomes sufficiently soft to be shaped by the blows of the hammer; but here it becomes perfectly fluid, and even rises so much above the melting-point as to admit of its being passed from the converter into a founder's ladle, and from thence to be transferred to several successive moulds.'

The metal thus produced was fine steel, and could be made for $6l.$ a ton, against something like $60l.$ a ton under the old system. The new Steel Age had indeed begun. Cheapness and rapidity were not the only recommendations of the new metal; it was, after a time, found to be superior also in quality to steel manufactured under the old system. We cannot follow here the history of the Bessemer process. It was so universally adopted that in 1868 it was bringing in to its inventor $100,000l.$ a year.[2]

Yet there was one great drawback to this system of steel-making. In the process just described one very common impurity of iron ores was not remedied, and that impurity was phosphorus. This was a matter of the highest practical importance; for the non-elimination of phosphorus rendered steel made in the converter from pig-iron containing it utterly useless, the phosphorus making the metal brittle and worthless. The result was that this wonderful invention could only be used for the conversion of pig-iron derived from non-phosphoric ores, and (since the

[2] Yet another mode of steel manufacture was a few years subsequently introduced: the 'Siemens-Martin or 'open hearth' process. It is not necessary in a book of this kind to describe this process. We duly note that it was subject to the same drawback, viz. non-dephosphorisation, as the Bessemer system, and that the 'Thomas-Gilchrist' process is equally applicable to it as we shall subsequently see.

old, long, and expensive 'puddling' process of Cort—in which the phosphorus *was* removed—could not compete on equal terms in the struggle with Bessemer), the great majority of British, French, German and Belgian ores became, to a large extent, unavailable for steel-making. In Great Britain the 'hematite' iron of Barrow-in-Furness speedily drove down in the market the phosphoric pig of Cleveland or of Wales; such pig falling or remaining stationary in price, while hematite doubled in value. The hematite iron ore to be found on the Continent (chiefly in Spain) was eagerly sought after.

How was it that phosphorus was retained in the Bessemer converter, and how could it be eliminated? If these questions could be answered satisfactorily—*i.e.* in such a way as to cheaply dephosphorise phosphoric pig—the cost of the production of steel could be again diminished, and the world would not only have begun its Steel Age, but definitely have broken with the Iron one. From 1860 onwards to the public announcement of the success of the Thomas-Gilchrist process, metallurgists were eagerly concerned with dephosphorisation. Sir Henry Bessemer himself, and an army of unsuccessful experimentalists, vainly grappled with the difficulty. Among other attempters of the adventure was Lowthian Bell, who had for years been regarded as the high priest of British metallurgy. In 1870-72 he published a work entitled, 'The Chemical Phenomena of Iron Smelting,' a book which must have been frequently in Thomas's hands. Doubtless Sidney had specially marked the following passage:—

'The limit to the production of Bessemer pig is want of ores free from phosphorus. The hematites of this country, under the sudden demand, have doubled in price, and speculators of all kinds are rushing off to Spain, where tracts of land, conceded without any payment a few months ago by the Government of that country, are said now to be worth large premiums; at least such is the impression left on the mind by a perusal of the published prospectuses of the day.

'This may be correct, and so firm may be the grip that phosphorus holds on iron, that breaking up the bonds that bind them together may defy the skill of our most scientific men; but it may be well to remember that the yearly make of iron from Cleveland stone alone contains about 30,000 tons of phosphorus, worth for agricultural purposes, were it in manure as phosphoric acid, above a quarter of a million, and that the money value difference between Cleveland and hematite iron is not short of four millions sterling, chiefly due to the presence of this 250,000*l.* worth of phosphorus.

'The Pattinson process does not leave one part of silver in 100,000 of lead; the Bessemer converter robs iron of almost every contamination except phosphorus, but nine-tenths of this ingredient is expelled by the puddling furnace. It may be difficult, but let it not be supposed that there would be any surprise excited in the minds of chemists if a simple and inexpensive process for separating iron and phosphorus were made known to-morrow, so that only one of the latter should be found in 5,000 of the former; and now that there is such a margin to stimulate exertion, we may be sure the minds of properly qualified persons will be directed towards the solution of a question of such national importance.'

CHAPTER V

YEARS OF EQUIPMENT

SUCH, then, was the problem Thomas had made up his mind
to solve. Of its solution, which was due to no sudden flash
of irradiating inspiration, but was the slow outcome of long
years of patient, tireless work, we will speak later. Its
consideration absorbed, month by month and year by year,
more of Thomas's scant leisure. After the summer of 1871
no more vacations were spent in mere voyaging for pleasure; every holiday was devoted in some way or other to
what had become the life object. The little laboratory he
had fitted up at home at The Grove became insufficient for
his needs, and he attended systematically the laboratories
of Mr. Chaloner (already mentioned) and of Mr. Vacher,
of Great Marlborough Street. He was determined, too,
to acquire all the credentials of the fully equipped practical
chemist, so that when the time came he might inspire
full confidence in men who would certainly doubt the
capability of a police-court clerk to overcome difficulties
which had baffled metallurgical chemists ever since the introduction of the Bessemer process. With this end in view,
he submitted himself from time to time to the Science
examinations of the Science and Art Department. From
obtaining the diploma of the School of Mines in Jermyn
Street he was excluded by the rule requiring attendance
at lectures; an attendance which he could not give so long
as the Thames Police Court claimed him; and the Thames

Police Court he was determined not to abandon until he had won for himself sure foothold and means of livelihood elsewhere. All the examinations at the School of Mines, however, which were open to him he passed.

We may mention here that more than one private friend, recognising Sidney's exceptional quality, and placing, perhaps, too much faith in the ' regular professions' as necessary to success in life, had offered Thomas some hundreds to spend in preparing for the Bar or Medicine. All such offers he had refused. In either case he must have abandoned his Civil Service certainty, since for ' walking the hospitals' his attendance at Arbour Square left him no time, and as for the Bar (although the preparation for that occupation is not of an arduous character), the regulations of the Inns of Court stood in the way, no clerk to magistrates being allowed to enter at those institutions.

On May 9, 1872, he passed at the School of Mines the examination in Mineralogy, ' first class advanced,' and on the same day in the following year the examination in Inorganic Chemistry, ' first class advanced.'

The summer holiday of 1872 was spent in Cornwall, the chief object of interest being the tin mines and ' works.' He travelled with Mr. Board, a fellow-student of chemistry. The pair had a letter of general introduction from Mr. Waddington Smyth, which enabled them to see much which would have been closed to unaccredited travellers. His mother and the rest had returned from abroad in the beginning of the year, and the old life at Camberwell Grove had been resumed; Sidney, with all his scientific studies and pursuits, with all his hard labour at his Court, being always the life and soul and central point of the home circle, never losing his interest either in domestic affairs or in more general questions of literature and life.

He began his 1873 holiday by accompanying his family
to Hythe, where he initiated his sister and brother into
geology; but he went thence to Bradford, whither he
was attracted by the meeting there in that year of the
British Association. Here we are enabled to quote again
the cousin who has described already the summer tour of
1869:—

'The four years since our French expedition had
ripened Sidney somewhat; yet in all essentials he was
the same, with his old keen relish for all intellectual things,
but with a rapidly intensifying bias towards practical
science, which was perceptible even to an outsider like
myself. In my father's house, where he was staying, the
visitors during the Association week were chiefly physiologists, and there was, I think, no one skilled in those
branches of knowledge which were becoming specially my
cousin's own. Yet he impressed everyone with whom he
came in contact with his exceptional acquirements and
ability—an impression which was certainly not marred
by the tact and modesty with which they were displayed.
That modesty he never lost, even after he had become
famous among all the metallurgists of the world. In
that, as in other things, he was genuine to the heart's
core of him; in all earnestness his own estimate of himself was ever too low rather than too high.

'As of old, many were our arguments together. One
of our chief battlefields was the vexed question of the
use of alcohol. The younger school of physiologists were
then in the first flush of the reaction against this dangerous
agent which has marked the medical history of the last
twenty years, a reaction which has now perhaps somewhat spent its force. Sidney, who personally had always
been almost a teetotaller, had seen much in his official
capacity of the devastating effects of the drink scourge,

and had gradually developed into an advocate of its legislative prohibition. I did not meet him (as in later years
I should have met him) by arguing that drunkenness was
a result of misery, and not a cause of it, but (being then a
fanatical partisan of personal rights and " Mill on Liberty ")
I went rather on the lines of the Bishop of Peterborough's
famous saying about " drunken freemen and sober slaves."
Starting from entirely opposite premises, we were thus
enabled to retain our own opinions, despite all contradic-
tion, with entire satisfaction to ourselves.

'Sidney took advantage of this visit to inspect the
famous Low Moor Ironworks. Together we attended
many of the sections, and I was more than ever impressed
with the wide range of his interest and knowledge. Yet
he was always ready to discuss the last novel of importance, even (if I pressed him) the last poem; although he
would still maintain his old heresy anent the superiority
of prose to verse. He teased me (I remember) by speaking slightingly of " The Earthly Paradise," as being in
truth unworthy of attention, since the book was no more
than it proclaimed itself—the work of the " idle singer of
an empty day." I discovered, however, that he had read
the " idle songs." '

It was out of this meeting that arose Thomas's first
contribution to 'Iron' (then edited by Mr. Chaloner),
' Letter on Bradford Hammers, and American Blowers.' [1]
From this time onwards for the next five or six years
Thomas was a regular contributor to this periodical. His
contributions range (as will be seen from the list printed
below) over a great variety of topics.[2] They were for the

[1] *Iron*, vol. ii. p. 712 (January 3, 1874).
[2] This list, which includes all
Thomas's articles in the first eleven
volumes of *Iron*, has been kindly
furnished to us by Mr. Chaloner.
Some six or eight small paragraphs
difficult to identify are excluded.
' He wrote,' says Mr. Chaloner,
' little or nothing in vol. xii.,

most part anonymous, 'but,' says Mr. Chaloner ('Iron,' July 6, 1885), 'his characteristic honour and rectitude appear in the fact that he never wrote a single line which would promote personal ends.'

Later in this year (1873), in November, Thomas was offered by Mr. Vallentine the post of analytical chemist to a great brewery at Burton-on-Trent, with a salary of 150*l*. a year to begin with.

This was through the kindness of Mr. Chaloner, already so often mentioned. The anti-alcoholic convictions which

which was the last under my care, and probably nothing but an occasional letter after that.'

'Bradford Hammers and American Blowers,' vol. ii. 712.

'Pollution of Rivers and its Prevention,' vol. ii. 771.

'Letter on the Refining and Converting Cast Iron,' vol. iv. 227.

'Metallurgical Text-books,' *ibid*.

'Heat without Coals,' *ibid*. 482.

'A New Philosophy,' *ibid*. 642.

'Current Thermics,' *ibid*. 674.

'Kinetics of the Future,' *ibid*. 802.

'A Budget of Heterodoxies,' v. 2.

'Oil Fuel,' *ibid*. 98.

'Coins and Coining,' *ibid*. 290, 355.

'Patent Cotton Gunpowder,' *ibid*. 162.

'Gun Cotton,' *ibid*. 259.

'Some Recent Developments in the Technology of Iron,' v. 290, 354, 418, 547; vi. 66, 418, 482, 578, 674, 771; vii. 67, 322.

'The Zinc Process for Lead Desilverising,' v. 424.

'Manufacture of Silesian Muffles,' *ibid*. 643.

'Percy's Metallurgy,' *ibid*. 706.

'Spectroscopic Estimation of Phosphorus in Iron and Steel,' *ibid*. 709.

'Historical Blast Furnaces,' vi. 4, 162, 323.

'A Gold Quest,' *ibid*. 194.

'Magnetism of Electricity,' *ibid*. 714.

'Charcoal-burning,' *ibid*. 802.

'A New Safety Tuyere,' *ibid*. 803.

'A Plea for Air Lines,' vii. 1.

'The Coming Air Lines,' *ibid*. 67.

'The Complete Bessemer Process,' *ibid*. 407.

'The Loan Collection of Scientific Apparatus,' *ibid*. 610.

'Recent Mining Literature, *ibid*. 770.

'Class-books of Chemistry,' viii. 34

'A Furnace of the Future' (first signed article), *ibid*. 354, 386, 419.

'Presidential Science,' *ibid*. 802.

'Technical Travel Talk,' vol. ix. 2, 66, 162, 258, 355, 451, 675; x. 2, 259, 451, 546, 674.

'The Swedish School of Mines' (qu. ?), xi. 98.

'A Policy for the Iron Trade,' *ibid*. 321.

'New Light on Steel-making,' *ibid*. 804.

This list alone would show Sidney Thomas's mental activity.

his cousin had noticed above had, however, by this time
become firmly fixed, and he felt that he could not conscientiously accept such a berth. Thus influenced, he
declined what in itself would have been to him a most
agreeable occupation, and continued his drudgery at the
Thames Police Court.

Early in 1874 we begin to be assisted in our narrative
by a series of letters (fortunately preserved) from Sidney
to his cousin Miss Burton, already spoken of. Miss Burton was now settled at Wiesbaden. We give here some
of these epistles belonging to this period:—

To Miss Burton
' 64 Camberwell Grove : March 20, 1874.

'Dear Bess, . . . You don't say if that wonderful
Kursaal supplies books as well as everything else, I mean
books as apart from periodicals. By-the-bye, I should not
go in for the *Leben Jesu* sort of literature. It will do you
no good, and unless you take up the whole question
earnestly and studiously, the impressions you derive from
it are valueless as conclusions, and to *you* particularly only
mischievous in their results. I don't send Latin Dictionary; why waste your time on Latin? Far better [spend
it] on German and Science. If you really want a Dictionary, you could get it better where you are, say in the
Tauchnitz edition. . . . For myself, since you ask it, I jog
on as usual. . . . I find more and more I *cannot* work as
I would, and doubt the wisdom of not giving self up to the
reverse. I certainly shall *after* June, if not before. It is
still drawing and struggling with pencils which no longer
have sharp points or any points at all. I wrote to " Iron " to
say I could not do anything in that line but *had* after all.
. . . I have no taste for the pen. . . . Have just spent an
evening with W. . . . We talked at a great rate on in-

numerable topics; disagreed on all, and he only resorted to
flat contradictions half-a-dozen times . . . Have been enjoying Huxley's "Lay Sermons," one at a time, enormously.
They bear a second reading; the ultimate test of a book.
Paget [3] has just published a volume of Essays, contributed
mostly to "Blackwood " . . . One on Ruskin and one on
Rubens at Antwerp particularly good.'

'64 Camberwell Grove : April 15, 1874.

'Dear Bess,—Went to a lecture at Society of Arts on
Friday, on a manufacturing subject; very interesting. If,
we were in town, I think I should go in for the Society.

'Nothing more suspicious about going to South Kensington than a wish to consult some books. I find the
library there as good for many purposes as the British.

'I think of going in for examination in drawing next
week. Though I fail, I shall have worked at a subject I
hate, in itself the best of educational processes.'

This examination was at the School of Mines—in
Applied Mechanics and Mechanical Drawing—and was
successfully passed.

In May 1874 he passed two further examinations at
the Science and Art Department: on May 1, in Steam,
'second class advanced,' and on the 25th, in Applied
Mechanics, 'first class advanced.'

The following short extracts from letters belong to
April and May of this year:—

To Miss Burton

'A re-reading of Trollope's "Australia" convinces
me that *Tasmania* is after all the ideal country, conjointly
with the South Sea Islands, and California perhaps.

---

[3] Mr. Paget, the Metropolitan Magistrate, who then presided at
Thames Police Court.

Everyone seems to concur in saying it is the most charming place for climate and productions in the world. Everyone seems to make his fortune in Ceylon.

'My friends the magistrates are exceedingly happy just now, having secured a long-sought extra 300l. a year. I am doing nothing now but a review of scientific basis &c. of iron-smelting, which means a great deal of voluminous reading with little result. Your account of your bird-pet delightful. Caged birds are an abomination, and the cat gets at uncaged.'

'For ten days I have absolutely and entirely been idle, and feel correspondingly despondent. All the rest of our small world lively in the extreme. A tempting offer came across me the other day of going to the South of France, but I could not afford it, as the salary but trifling. I long for change.'

In the summer of this year the household transferred itself to Sussex Place, South Kensington, where for the next three years the family dwelt. The next letter is dated from the new abode:—

'18 Sussex Place, Onslow Square, S.W.: 1874.

'Dear Bess,—I returned just in time to go [up] for the last examination I have in view before settling down to a peaceful and indolent old age, with what result I know not, but will not post this till I do.

'Since then we have been in a whirl of move, move, packing and packing, than which nothing can be more abominable. Heaven defend me from being possessed of any chattels of my own.

'As I have been pretty regularly tied to the Thames till 6.30 or 7, I am beginning again to consider how excellent a thing is rest. My chief solace has been Mill's "Autobiography;" it is quite a pearl amongst books, earnest, thoughtful, and carrying a conviction of entire candour. Our present nearness to the [South Kensington] Museum Library will be a great boon, though one cannot take books out.

'Your life at Baden seems a very bright one. I suppose it is, as you say, just the life to suit you. I myself sometimes feel quite a desire for foreign scenes and manners.

'Lil and I went to a spiritualistic *séance* at V——'s shortly since; two lady cousins of his, a mutual friend and ourselves, forming with the medium the "circle." Though it was not considered a satisfactory performance, I saw several matters which I have as yet failed to find an explanation of. V—— himself is a red-hot convert, and is now firmly convinced of immortality, having been previously a gross materialist.

'Was at the "Throat and Ear" last night. The infirmities of humanity, as seen at any hospital, form anything but a cheering spectacle, and I came away depressed; though Llewe [4] was very nice, and anxious to display foul depths of his patients' throats and ears with the most picturesque light of healing science.

'I start on August 3 for South Wales.'

His usual holiday this year was spent partly with his cousin, Mr. Percy Gilchrist, then chemist to the Cwm Avon works in Glamorganshire, partly at the British Association meeting in Belfast, and partly at Bradford. The following extracts from letters describe it sufficiently :—

To Miss Burton

'Glamorganshire: August 3, 1874.

'Dear Bess,—I have at last started fair on my holiday-making, though I feel it rather selfish to leave the mother

[4] The late Dr. Llewellyn Thomas, Sidney's elder brother.

and Lil at home. I [am] so glad to get away. My last
month not overworked and worried. By-the-bye, I *did*
fall through both the final examinations I went [in] for,
though I have no particular gratification thereat now that
it is done. I had rather an amusing occupation lately—the
correction of a translation of a French pamphlet! The
idea of *my* correcting any translation I regard as rich
in the extreme. However, as it was a technical subject, I
was able to earn quite a reputation as a French scholar.'

To his Mother

My only excurse has been to Siemens's Works,[5] where I
spent five hours; came out looking like a stoker, and was
thrice drowned coming back, all of which I enjoyed.

' When I go to works we generally go up in a superb
passenger car which tails on to the trucks always in transit
'twixt harbour and works.

' I shall probably go to Belfast on Monday or Tuesday,
but will let you know before I start. I feel it dreadfully
selfish for me to be down here; should so enjoy having
you and Lil with me.—Affectionately yours,

'S. G. T.'

Belfast, 1874.

' Dearest Mother,—I have just got your letter; very
glad to do so. Chaloner is here in great force. I am with
him a good deal, as he knows several amusing characters,
an *Hour* man . . . great fun, several other pressmen, and
others. Went with him yesterday to Giant's Causeway, a
dreadful railway journey, but magnificent cliff scenery; not
quite up to one's expectations possibly; but that is human

---

[5] Thomas had been given by Mr.
Walter White (the late Assistant
Secretary to the Royal Society) a
letter of introduction to Sir W.
Siemens.

nature, or my nature at all events. We walked half way from the nearest station, and then had a boat along the coast, which I enjoyed immensely. On Saturday Odling's lecture was a treat.

'I quite look forward to seeing you.'

To Miss Burton

'You will have read of the sayings and doings of the associated savants. The two lectures of Huxley and Lubbock you should not miss on any account. They were reported in the *Times*, which I understand you see. Tyndall's address, eloquent though it was, was hardly to my mind satisfactory.'

Back in town, and now at Sussex Place, the routine of his 'double life' was little changed. Only, instead of walking the whole way to the court, as had been his practice in Camberwell, he would take train to the City and thence tramp to Arbour Square. He was now systematically working at dephosphorisation and gradually feeling his way to a solution.

The following letter tells something of Thomas's not too numerous recreations :—

To Miss Burton

'November 21, 1874.

'Dear Bess,—I was taken to an Albert Hall concert last night and heard Von Bulow play marvellous tricks with the piano; *tours de force* they seemed to my unenlightened mind. (How is your music going ?) The Hall looks magnificent, but it is not half filled. They are trying concerts every night, and the Briton soon wearies.

'I have done a few articles for " Iron " lately, but only regard it as education. It is not my *forte* (if I have any), and takes up too much time to pay. I am obliged to husband my health resources, I find, after all.

'I had a pleasant little dinner at V——'s shortly since. He had what I regard as the infinite good taste and sense to ask three or four men only and provide an entirely simple meal, such as he would have by himself. An old assistant of his has recently returned from Servia, which appears a virgin country, ripe for the most profitable exploitation. It costs about 20*l.* a year to live *en prince*, with gold and silver and lead and forests of finest timber to work on. Three English capitalists have gone out to found a little state, starting with a few hundred square miles. V—— is quite a pet of the mother's. His spiritualism is a little coming down.

'You will have heard of the immense success of Farrar's "Life of Christ." Some one has insisted on lending it me. I like the preface. You should read it if you can. What is wanted now is an answer to "Supernatural Religion" by a man at once able, erudite and wide-viewed, answering it on its own ground and not on quite another platform; and then the world may decide on adequate grounds on the most momentous of all questions. Does "Nature" penetrate to Wiesbaden? It boasts an European circulation and gives shortly a sketch of current science. I have a dreadful budget of things from Chaloner[6] he wants me to make something of. . . . I have only seen abstracts of Gladstone's pamphlets. He has, at all events, brought out a latent Old Catholic party in England.—Yours,

'S. G. T.'

Early in the following year of 1875 we find Thomas again writing to his Wiesbaden correspondent :—

[6] See 'A Budget of Heterodoxies,' *Iron,* v. 2.

To Miss Burton

'18 Sussex Place, Queen's Gate, Kensington :
'March 18, 1875.

'Your note just received starts me on my epistolary labours, which I should otherwise have attacked very shortly. It is pleasant to hear of your being in high spirits.

'I shall certainly try to look you up this summer, but, if the mountain will not come to Mahomet, Mahomet must come to the mountain, which is at present located at South Kensington ; where its site will be in the autumn I know not ; we have to settle shortly whether we stay here.

'I am over ears in a technical experimental investigation on Iron which is likely to last me considerably, and then perhaps to have no result ; but, after all, life is very little else but the pursuit of crotchets, the pursuit being the best part of it. I recreated myself after a long spell at references by a rink yesterday. I had not been for some time, and found the wheels more popular than ever. The elaboration of costuming it has developed is quite a phenomenon. Do you read the English papers ? I understand you have access to them. You ought not to allow yourself to become behindhand in the manners and customs and literature of your native land. I shall submit you to an examination thereon when we meet.—Yours,

'S. G. T.'

Of course the 'crotchet,' so lightly spoken of, was dephosphorisation, the solution of which question was now beginning to assume shape and consistency in Thomas's brain.

The next letter is one of thanks for some birthday present, and incidentally expresses certain humorously distorted views of the German language and people :—

To Miss Burton

'Sussex Place: April 17, 1875.

'Dear Bess,—Your good wishes, which reached me yesterday, pleasant to receive and appreciated; though my theoretic objections to presents are, you know, profound, I also appreciate and thank you very much for the pleasant and practical and most useful token of remembrance you caused to be conveyed to me. I was, in fact, only waiting till after the 16th was past to ask you to get me a technological dictionary. Your idea of my German scholarship is delightful. Do you know it took me half an hour to translate the first ten lines of the cutting you sent me, and then I was not clear about them? I consider, if I don't have to look out more than two words in a line, it is a special providence. As for the Germans, I consider that their existence on this earth, taken in connection with their barbarous, unintelligible, cumbrous, inelegant and never to-be-sufficiently-deprecated so-called language, is a blot and stain on the fair reputation of this continent. I have pleasure in observing similar sentiments pervade the appreciative periodical writers to whom you allude. Your views appear to have been slightly modified by your pleasant surroundings, but you will doubtless agree that the independent and impartial opinion of the insular observer is most calculated to come to a correct conclusion.

'I have some idea of getting up a little elementary Spanish.'

The next letter seems written under the impression of some temporary check to the dephosphorisation investigation.

To Miss Burton

'Thames Police Court: May 15 [1875].

'Dear Bess,—My blunder shows the difficulty of combining the inconsistent occupations of note-taking, with the

E

innumerable distractions under which it is performed, and letter-writing. I am afraid my "Iron" contributions would be hopelessly uninviting to you, or I should send them, but mere "iron," "heat," "furnaces" and so on would be an imposition on you. I went the other day to private view of the Scientific Apparatus Exhibition at South Kensington, and was greatly surprised at its extent and interest; it is one of the best strokes for science that an English department has yet achieved. You are to be envied if it were only for adjacent woods. It is pleasant to think of your being so happily located. As for London, bah!

'I am all behindhand with work both here and at home, with a pile of books to review. I have been spending much time and labour over an investigation which has not resulted in anything useful, and am considerably knocked up, not to say ultra seedy.—Yours,

'S. G. T.'

The holiday this year was spent in Wales, and not in Germany, as had been hoped; visits to 'Works' alternating with long tramps.

The following letter tells us something of Thomas's movements :—

To Miss Burton

'B——: Sunday.

'Dear Bess,—I walked over here from Neath. Have been here since Tuesday, and am off again to-morrow. I am with a man I have some slight acquaintance with who is engaged at some works at B——; not a very lively place, though on the sea; and with a small dock, about a mile of sandy flats 'twixt hills and sea. Three large metal works and that is all. I amuse myself as best I can 'twixt hills and sea. I have some idea of a two days' ramble in the interior, then looking in on Percy's home. It doesn't come up by a long way to my anticipated German holiday, but is the best I can manage.

'Now I have some assistance to ask of you. It is this : Would you get Stumner's "Ingenieur" (published Vienna) for June 18, 1875, through a bookseller or direct? In it is the continuation of an article " Hochofen, Anlage auf, &c.— Gleiwitz." I would send the paper, but it is mislaid. I am making a summarised translation of the set; and it would be of great service to me if you could give me a literal translation of that number (leaving out any words that are quite unknown to you) and send it with original to me, " Care of P. C. G., Cwm Avon."

'If it would weary or trouble you don't think more of it.'

It is right to mention that these letters to Miss Burton are filled with information and advice about investments and finance, advice which it has not been thought necessary to reproduce. As we have said above, Thomas, amid all his ' numerous and engrossing occupations, found time in some mysterious way to conduct the affairs of more than one lady relative.

Here are two letters written about this time to his sister Lilian (then at school at Richmond) which show something of what may be called the domestic side of the character of Thomas.

To his Sister.

' Dearest Little Woman,—Sentiments of the most profound satisfaction inspire the fraternal breast at the tidings of the moral and intellectual reformation which has taken place since you left me, dissolved in tears, on the South Kensington platform. All hail! O taciturn, virtuously at 6 A.M. arising, and much fasting sister !

E 2

Fail not in thy praiseworthy career, and receive a double first class Local Cam., Oxford and London University degree, with accumulated honours in the natural sciences, notably in your favourite pursuit of chemistry.

'To return to things sublunary. Grind muchly at *German*. I have undertaken to do (or get done) another German translation of prodigious dimensions and unutterable obscurity, solely with a view to keep up my imaginary reputation for translatory capacity, so that I may shift it to your juvenile and competent shoulders, as a step towards a pleasanter independence than the scholastic.

'Needless to say that mother's bulletin chronicles minutely everything that does or does not occur *chez* No. 18. The only event is Llewe's doctorate at Brussels, which seems to have been gained with brilliant distinction and with compliments on his facility in French. I shall be off holiday-making on Saturday fortnight. I may possibly look you up the Thursday before I start, and if so, and you are very good, you shall have a row (you row and I steer). We won't dine at the " Star and Garter," it might make the rest jealous; but we will discourse sweet Chemistry instead.—Respectfully and affectionately,

'Your Brother.'

'The Eve of *the* Birthday : September 11, 1875.

'Dearest Little Maid,—Let me, with due submission and humility of mind, offer my fraternal felicitations to one who has reached the dizzy altitudes of antiquity to which your ladyship has scrambled. May the eventful 12th always pleasantly mark a step (or several) towards that culminating day on which I may see you as good and nice a little woman as I could wish you to be (which is equivalent to wishing you a few centuries of progressive existence). Enclosed a pair of prodigious wash-leather gauntlets, selected by the mother as suitable to your age (and destructive habits). I had contemplated a daintier pair ; but the perplexing question as to whether seventeen or one was the proper size hindered my venture. . . . In haste, and with love, your brother,

' SID.'

' So sorry you will not be with us, but you are quite right not to come. *Work !* '

Later in the year come some more letters to Miss Burton :—

To Miss Burton

' October 5, 1875.

'Dear Bess,—I, like you, not feeling remarkably brilliant ; still send a technical paper to " Iron " every few weeks, though I have no enthusiasm for that species of employment. I have been seeing something of a rarity—a student *bonâ fide* who learns languages *pour passer le temps*, and lives in a very pleasant studious retirement with that intent. I have been reading Matt. Arnold on Prussian education system, which certainly reads as approaching perfection, a view which our Teuton professor endorses. The *Times* in recent articles on their army, exhibits well the causes of their military superiority. The " Turkish question " not long since promised to afford an opportunity for a general European squabble. Chesney in " Macmillan " has proved to his satisfaction that Prussia and Russia are to be the next pair in the cockpit.'

In the next letter, already in 1875, and not then for the first time, a warning note is struck as to health :—

To Miss Burton

' November 1875.

' Dear Bess,—I hope to make sure of seeing you *chez vous* in the summer, unless any unforeseen event should intervene. I feel, however, slightly dubious as to my successful progress, as I have absolutely *no* German, my good resolutions in that direction having been interrupted. "Iron" now offers me as much work as I can do, but as the subjects I select require much reading, it is not remunerative. I am constantly "knocking up," a weakness to which I imagine I shall some day "cave in," unless I throw England up altogether.

'I should have sent you some "Irons" for criticism, but as my last eight or nine articles have been on Blast Furnaces I am not merciless enough to ask you to read them. What do you think of the *World*? It has made a great hit. Sells 39,000 a week. It started with a trifling capital, on which it pays a few 100 per cent. . . . *A propos* of art, of course you know Henschel's sketches in the photos; some are delicious. If I get time I will write more, but I have a book on charcoal, another on electricity, and two articles which I ought to be attacking.'

'December 15, 1875.

'Dear Bess,—. . . An American girl-student—pretty, too—has been visiting London hospitals, and to the disgrace of the students thereat has been insultingly warned off. She called at Llewe's hospital, where, of course, she was received politely.

'The Suez question is the great subject of discussion; all enthusiasm at first, but now a growing feeling of hesitancy about its benefits has supervened. The idle world is frantic on skating-rinks; they are springing up everywhere, and are crowded at all times. Have you one about Wiesbaden? Among a skating people like the Germans it would be a great success, both with natives and foreigners.

'December 22.

'I have kept this back so as to make it a Xmas letter. To my great comfort we are not going to have any Xmas festivities or visitors of any kind. My namesake of Bremerhaven is the most interesting problem that has ever been presented to the analytical moralist. In every relation of life he appears to have been perfect in amiability and *savoir faire*, exceptionally so, and yet throughout planning and carrying out the most infernal, deliberate, wholesale murder. A magnificent hero for a morbid psychological novelist. The man who wrote a startling book on New Guinea, which you mentioned was discredited in Germany, is by no means accepted here except as a modern Munchausen. I have asked you repeatedly what you *do* all day and every day.

'I send a new version of "Faust," the sketches in which may amuse you. With all good wishes for Xmas, and above all for 1876 and its successors, which I trust may bring you all happiness,—Yours,

'SIDNEY.

CHAPTER VI

THE PROBLEM THEORETICALLY SOLVED—A GERMAN TOUR

In the latter end of 1875 the great problem was approaching to, at any rate, provisional and theoretic solution in the mind of Thomas. He had gathered together all available analytical and technical *data*. The first question to be answered was obviously (as we have said above) the fundamental one—*why* was phosphorus retained in the Bessemer converter? That preliminary difficulty surmounted, the path might or might not be clear to cheap elimination; at any rate it would at least be visible.

Thomas came to the conclusion that the reason of the non-elimination of the phosphorus was to be sought in the chemical nature of the *lining* of the Bessemer converter. This lining has been described above in Sir Henry Bessemer's own words; it varied in material, but the material, whatever it might be, was acid in chemical essence. The phosphorus in the iron was rapidly oxidised during the process, or, in other words, formed phosphoric acid. With an acid lining that phosphoric acid would not combine, the two acids having no 'chemical affinity' or liking for each other.

If this were the cause of non-elimination, the path to be followed was visible indeed. Not by any addition or mixture of substances after the converter had been charged was solution to be found, but rather by a change in the constitution of the lining. For the acid lining in use a *basic* one must be substituted. A base is a term used by chemists to signify a substance which will combine with an acid, a substance for which an acid has ' affinity.' Some strong base then must be employed for the lining.

Thomas entered upon a series of experiments for the purpose of investigating the material and duration of various linings. Durability was essential to cheapness and, therefore, to commercial success, and a substance which would long survive the intense heat of the Bessemer process was by no means easy to find. Thomas at this time came to the conclusion that the required material must be either lime or its congeners, magnesia, magnesian limestone, &c.

It must be remembered always that the aim to be attained was twofold, as will be seen by the quotation from Lowthian Bell, *ante*, p. 34. Perhaps the more important object was to separate the phosphorus from the iron ; but it was also of great importance to preserve the phosphorus, which (noxious as it was when combined with iron) was in itself a most valuable product, at least in the form of phosphoric acid. This could be done by creating a basic ' slag.'

So far, then, had theorising and experiment led Thomas at the end of 1875. He was convinced that his conclusions were chemically correct, but he found it impossible to finally verify them under such conditions as were open to him in his rough little laboratory. He attempted in his top room at Sussex Place to obtain a Bessemer blow by means of an improvised converter in the ordinary domestic firegrate, which was alone at his disposal ; but he naturally found it impossible to obtain the necessary blast.

Thomas thought, however, that he saw his way to more satisfactory trial of his theories. A cousin, Mr. P. C. Gilchrist, already mentioned, was, as we have seen, then

chemist to certain Works at Cwm Avon, in South Wales.
It might be that Gilchrist, although, of course, he had no
unlimited command of the works and appliances, might
at least be in a position to experimentalise more satisfactorily than was possible in Sussex Place. Early in 1876
Thomas wrote to him communicating his theory in detail,
as well as the lines on which he thought it could be proved
or disproved. Gilchrist at first deemed the whole thing
a chimera, but undertook, nevertheless, to make some
experiments. The business, however, slumbered for long
months; Thomas on his side still working at his idea in
the evenings at home and devising the best method and
the best materials to make the experiments a success. In
the summer of this year we find him writing to Gilchrist
under date of August 7, 1876, from the Thames Police
Court :—

'My impression is, a biggish wrought-iron crucible
would be as good for experimental converter as anything,
and would be easy to try various linings in. The tuyeres,[1]
subject to your emendations, might be pieces of wrought-
iron gas-pipe covered with fire-clay and with fire-clay
stopper perforated thus — or laterally. I have not time
enough to *do*. I only go home to sleep and eat. Most
unsatisfactory.'

For some months yet, however, Sidney had to continue
to chafe at delay.

Meanwhile he had found time for a July holiday in
Germany, a holiday mainly spent in visiting Works. The
following letters to his Wiesbaden correspondent were
written before, during and after this time :—

[1] These, it will be remembered, are the air-holes of the converter. (See Sir H. Bessemer's description of his process, *ante*, p. 32).

To Miss Burton

'18 Sussex Place : June 1876.

'Dear Bess,—Plunged over head and ears in work. I look forward to starting to your beloved Germany on Monday night, the 3rd prox., if I can find time before then to address myself to the necessary consultations of Bradshaw, &c., provided always that the mother is well enough to get away to the sea without me. Now, though my bourne is the Hartz, I need hardly say I contemplate being in Wiesbaden, if not *en route* at least on my homeward voyage, that is, if you care to see me. So I want you to write when you will prefer my going, beginning or end of July. I have a man who talks of accompanying me, but I shall probably be alone. All news such as there is may be best delivered orally. I mean to travel without any luggage but a pen and an umbrella, a hat and a dictionary. Will you be shocked at the introduction of so uncouth a traveller amid the refinements of Wiesbaden ?—Yours,

'S. G. T.'

'Dear Bess,—I formulated three conclusions before my arrival at Frankfort :

'That I am very sorry I have come to you first and not last, as I had intended, on the principle of keeping the pleasantest of everything to the last.

'That I would try to bring my holiday in your direction next year.

'That if I had stayed a day longer the Hartz Bergwerke &c. would have been shelved altogether. From which reflections (added to one that I had not said half I intended), I was aroused by arrival at Frankfort, which I proceeded to *do* in the time I had to spare. I will not trouble you with any hasty observations thereon. The seven hours to Eisenach were tedious, though the country somewhat interesting; more so my fellow-travellers, especially a young soldier and an artist, the latter just returned from a sketching excursion in Schweitz. These with two others kept up a lively interchange of jokes and information. I, a silent spectator, could only catch one-fifth of the points.

'At Eisenach we parted; the soldier gave us all his name and address, and we him our cards. Hope he won't call and borrow.

'At Eisenach to a good hotel, and was off by 6 A.M. to Wartburg, which I accomplished with a party of students. Then through rain "fahrers" to Austhal, which I happily stumbled on at one. Both Burg and Thal sehr romantisch and so on. Dann hat ein teuflich Fahrer mir misdirected, und habe ich zwei Stunde aus von mein Weg gegangen. Then through forest to Rluhla, a curious miniature Bad with Curhaus, and so on in a hill valley; on again through woods and over hills to a primitive Dorf, where I put up at a primitive hostel with a getrunken Wirth wer zu mir Deutsch sprechen insisted. My bedroom, shared with a Fuhrmann, though deficient in some elegancies, was ziemlich bequem. Morgens früh über Friedrichroda another Bad, nach Oberhof, on the way picking up a student. The infamous Schurke had on me his infamous fraud perpetrated; he said he Englisch konnte, aber Englisch kann er kein Wort verstehen. Through a beautiful rocky valley, up a series of hills, and then twelve miles of continuous wood, brought us to a Gasthaus, wo ich ein wunderbar Milchkur habe gemacht.

'Morgen früh nach Ilmenau by Berliners frequented Wasser-Kur und Austall wo ich mit meiner Student with much vergnügen parted. Then to Konigsee; curious old town, excessively hot, so I in a hasty Augenblick der *Post* genommen habe. Der Post a wicked snare and vile delu-

sion, kann ein Meile in ein Stunde ; and as for the horses—
Donner Blitz!

'A postman entered into conversation with me, and
gave me a commission to execute in London with mystic
names and so on, on paper. I don't know what it was I
undertook, but we parted great friends. Half way to
Rudolstadt my post got emptied, and Kutscher wanted
me to ansteigen, which I declined to do, having my billet
further genommen. I argued the question in my native
tongue, and utterly routed Herr geehrter Kutscher. An
appalling nine hours' train to Chemnitz, where I got at
10.30. Asked a young person with a brilliant cap to direct
me to a Gasthaus, and after er hat das gethan, he insisted
on drinking beer and talking German to me till 12.30.
Oh, horrors! what I suffered with him! also exchanged
cards, swore eternal friendship, and so on. I wondered
what he said all those two hours. I said So? Ja! Ja!
So? which satisfied him.

'Morgen früh nach Freiberg, wo ich bin, got a fair on ;
queer place. I have been much longer getting here than
I calculated. In Thüringen Wald, to get five miles in a
straight line, you had to go eighteen.

'I shall not go to Essen now. It is quite possible that
Herr F. may also not care to have strangers on his works.
I should like to know if this be so early. Would you
write me a card both to Mansfeld and Thale Hartz as
to this, and send my bag to Kreimsen? Shall be in
Dresden Tuesday ; no time for Saxon Schweitz.

'The only German who can speak English, I believe,
lives on the Rhine. We must push on the universal tongue.'

'Dear Bess,—Here I am at the end of my tether, and
preparing for stringent harness. I received yours and

cards (for which many thanks to both of you) at Clausthal
and Goslar. In case you interest yourself in my remaining travels, here they are. From Mansfeld, whence I
wrote you, and where I accomplished some works, I pere-
grinated to Hartz Gerode. Uninteresting works, hot and
dusty. H——e nothing to boast of, but so-called castle
sleepy and primitive. Thence to Alexisbon, another
miniature Bad, buried in a valley, woods all round, a dirty
stream, said to be irony, and salubrious Band Curhaus,
and frequent refreshments. So over a hill through a wood
to a schoenes Aussicht. Had to climb up a tower—my
tenth—where a ruffian persisted in showing off his topo-
graphical lore by pointing out to me every village within
the horizon. Again to Rosstroppe and Tanzplatz—really
a fine view—where all the cits of North Germania were
drinking and singing to their great content; sleeping at
Thale; on again by Blankenberg, striking the Bodathal
again at Rubeland—last again pretty—and halting at
Elbingerode. Hence a lovely walk in early morn through
woods up Brocken, whence I gazed my fill and lighted on
a delightful little sylvan inn by Andreasberg. Going
down a mine and over works at Andreasberg, which is
also now frequented by "fir needle" bathers, occupied most
of next day. My next stage Clausthal, where I stumbled
on a Londoner—University student—with whom I did the
"Lione," escorted by two German students. So round
Ochretal and on again to quaint old Goslar, and on again
to Kreimsen, where I picked up my bag. By train to
Mulham near Ruhrort, and by seven on Monday morning I had the audacity to call on Herr. Dr. F., whom I
found at breakfast with Mrs. F. and an amusing young
lady of two. Was received most courteously, and taken
to Phœnix, where I was left to satisfy my curiosity, which
I did at length, finding the works well constructed and

worked. I was to see Herr F. again, but unfortunately he did not return to his office before I was obliged to leave to catch the only train to Rotterdam. I left a card expressing my thanks. There are several points on which I may possibly write to him for information. Does the director read English I wonder?—Yours ever,

'S. G. T.

' P.S.—My opinion of German scenery is—is reserved ; of the folk I can say I have a much better opinion than I started with. If they would only learn English they would be civilised.'

'Dear Bess,—Here everything going much as usual. My editorial acquaintance just back from America; speaking well of things American, particularly of their extraordinary capacity for work and rapidity in executing it. Awaiting my return I found a letter from my friend in the Western States saying that he was relinquishing the Professorship he has hitherto held, and suggesting I should take his place. It was a temptation ; but, of course, in my mother's state of health it would have been out of the question.

'I find so much to engage me that it is doubtful whether I shall have time to turn my German visit to any literary account, particularly as a great part of my notes got lost in hurry to catch a train for Ruhrort.

'By the way, as to "hurry," you seem to think my time is unlimited ; I had twenty-six days for all.

'I enclose a number of queries, of which the director may answer *some* in German or English possibly, if you would kindly undertake their transmission. They are simply what I had jotted down at the time to ask the director before I left. Of course it is a considerable trespass, on the strength of your introduction ; but I find

German scientists so courteous in giving information that
I have become a hardened interrogator.'

'Dear Bess,—I am intensely obliged to all of you, the
Doctor, Fräulein N. and yourself, for the trouble you have
taken over my troublesome interrogatories, which I certainly did not expect to get *so* answered. You say that
Phœnix had forty-eight furnaces at work in 1872–73, now
only eighteen. Does that mean blast furnaces (Hohofen)?
for if so, Phœnix is larger than I imagined; few English
works have more than twenty in all. By asking the *name*
of the hot-blast stoves I meant this: I observed in particular one new hot-blast stove (*i.e.* an apparatus for heat-
ing the blast before it enters the Hohofen) of a construction
new to me. I know the Whitwell stove, the Cowper, the
Pistop pipe stove and so on. This appeared to be filled
with circular discs of iron (?), so I asked by what name it
is known that I might find a description of it. *In einzeln*
etc. means "is more tenacious." *Hartenummern* I should
translate as "scale of hardness" I fancy, but I am not
quite clear; what is your idea? The director's answers are
admirably clear and to the point.

'I will send "Iron" to Herr Dr. F. as you suggest.
It is simply appallingly hot, and I find Thames has
effectually taken all the good I derived from my trip to
itself. The amount of work accumulated is quite a feature,
and I have a new magistrate. Wish I could exchange
Kensington for Wiesbaden for a week or two.—Yours,

'SID.'

CHAPTER VII

'TECHNICAL TRAVEL TALK'

THOMAS *did* 'turn his German tour to literary account' by
the contribution of a series of articles (under the heading
of 'Technical Travel Talk') to the columns of 'Iron.' We
reproduce some extracts from these articles (published in
the course of 1877) here. Much of them is, of course, too
technical for these pages. The opening paragraph is very
characteristic of the writer :—

'Freiberg.

'There is a curious delusion very prevalent among
vacation-tourists, that it is inconsistent with the purpose
of true holiday-making, and indicative of a certain poverty
of spirit, to concern oneself about aught else than the
picturesque and artistic features of one's holiday-ground.
By such a limited interpretation of the available resources
of pleasure-travel, not a few are condemned to hours of
ennui, which they would escape effectually if they would
only recognise that the industries and institutions of
a strange locality are as legitimate objects of interest as its
scenery, buildings and pictures. Of course there are those
who are so profoundly convinced that instruction and
amusement are hopelessly incompatible, that they are
consistent in refusing to desert the beaten tourist track,
lest perchance they should fall into the pitfall of instruction. It cannot, however, be believed that, of the thou-

F

sands of Englishmen who sojourn in or pass through
Dresden yearly, all labour under this singular prejudice,
and believe that it is incumbent on a true holiday-maker
to utterly bury and forget all the interests which constitute
the chief concern of his everyday life. Yet it is surprising how few of our practical countrymen find their way
from the art-capital of Germany to the old mine-city of
Freiberg, the birthplace of technical education, and of the
systematic application of scientific methods to the conduct
of industrial enterprise, though the two places are barely
an hour's ride apart.

'The district of the Saxon Erzgebirge (Ore-mountains),
of which Freiberg is the centre, would, indeed, be well
worth a visit, even though its only attractions were the
quaint and picturesque architecture of its towns and the
primitive customs of its people, among whom the eerie
superstitions and legends, which filled so important a part
in the lives of the old miners, still linger.

'Freiberg itself has seen fluctuations of fortune beyond
the experience of ordinary cities. To have been the scene
of many sieges, the cradle of the Saxon Reformation, and
the seat and city of refuge of the royal family of Saxony
are only a few incidents in its chequered political career.
Its real prosperity, however, fluctuated with that of the
mines of the district, and the depreciation and appreciation of silver was a question of deep moment to its
burghers long before the dwellers in Lombard Street had
begun to dabble in the intricacies of finance. In the
sixteenth century, when its mines were at their best, the
population of the city is said to have been five times as
great as it was at the beginning of the present century,
and considerably larger than it is at present.

'The contrast between the mediæval streets and fantastic buildings of the old town, and the costumes and

manners of the crowds that thronged them was particularly striking as I made my way from the station and found the Jahrzeit, or semi-annual fair, in full swing, with all the accompaniments of bands, shows, jugglers and vociferous cheap-jacks. Strolling through the good-humoured multitude I came on a little group of American academy students, who were laughingly engaged in showing the heathens (as they designated the non-English-speaking portion of the community), in some trials of strength, that transatlantic skill could prevail over Saxon muscle. High over the busiest part of the fair loomed a mining engine-house, perched on the inevitable rubbish mound, requiring no great stretch of the imagination to picture it as the genius of the place. The monotonous periodical clang of the engine-bell, which throughout the mining region serves to indicate that the pumping machinery is in order and at work, readily lends itself to this fancy, by giving to the stranger an almost painful consciousness of automatic, never-tiring watchfulness.

. . . . . .

'As some salt carriers from Halle were making their way across the Freiberg heights with their salt, on their way to Bohemia, it chanced that one of them picked up by the roadside a lump of lead ore. Being evidently shrewd and enterprising men, they abandoned their Bohemian journey and betook themselves with their find to an eminent assayer at Goslar. A certificate having been obtained that their specimen assayed much richer in silver than the ordinary Rammelsberg ores, the fortunes of Freiberg were made, for divers Goslarites emigrated forthwith, and speedily opened up the rich silver deposits which soon rendered Freiberg one of the most prosperous cities of Central Europe. What became of the original enterprising prospectors, Agricola, who is the authority for this account, does not chronicle. The author of a curious little work on "The Origin of the Saxon Mines," published at Chemnitz in 1764, discusses the question of the exact date of this discovery in great detail, but if we follow Agricola again in fixing it in 1164 we shall not be far wrong. Between the years 1164 and 1824 the Saxon mines are said to have produced 4,100 tons of silver, valued at thirty-six millions sterling. Their greatest productiveness appears to have been reached in the fourteenth and fifteenth centuries, when there can be no doubt that some of the richest veins were struck and almost exhausted, large masses of ore, yielding sixty and seventy per cent. of silver, being found.

'In 1810 the product of the Saxon silver mines was estimated at 53,000 marks, or, say, one-eighth of a million sterling. In 1817 it had sunk to a considerably lower value. In 1850 we find it still at about the same figure, though the total value of the mineral products of Saxony had doubled in the interval. In 1856, however, the production amounted to 55,000 lb. of metal, and in 1865 to 80,000 lb., while by the last returns from the Freiberg smelting works the value of the silver produced has again declined.

' At the date of the last official return there were in existence, in the four Reviere into which the ore-mining district of Saxony is divided, 344 mines. In this numeration, however, are included drainage and extraction adits, and over 150 mines which are not in work at all. Of the balance, only nine were in the dividend list, while sixty-four of those reckoned as " going concerns " were raising no ore. The total ore raised in 1874 amounted to about 50,000 tons, representing a cash value of something over 250,000l. sterling. Of the 76,000l., which was the value of the ore raised from Himmelfahrt, the most prosperous of all the mines, only 11,000*l*. went into the pockets of the shareholders.

'The Himmelfurst mine at Brand, some two miles or more from Freiberg, is one of the most important in the district after Himmelfahrt, which is the show-mine to which visitors are usually directed, and where there is accordingly less opportunity of seeing the normal course of mining operations than elsewhere. Soon after five on a rainy morning I met, by appointment, in the Freiberg market-place, a figure clad in coarse miner's dress, patched from top to toe with earth stains, and duly adorned with leathern apron and belt, a knife and a lamp. This costume is the regular mining costume of Saxony, where miners dress, not, as is the wont at home, as individual taste or convenience suggests, but just as their fathers and fore-fathers did before them. The wearer, however, is an English student, a chance acquaintance, to whose courtesy and intelligence I was much indebted. After a wet trudge along an elevated highroad, bordered by a monotonous country, which, hedgeless and almost treeless, looked bleak enough even in summer-time, and recalled the fact that agriculture in the Saxon uplands is a precarious pursuit, we arrived at our destination. At intervals along the road we had exchanged a friendly " Glück auf," the universal salutation for all times and occasions in mining Germany, with individuals accoutred like my companion, hurrying to their respective mines; but as we entered the group of offices " Glück auf" is heard on all sides. My friend having interviewed the presiding official and shown his academical voucher, and the usual preliminary of entering our names, domiciles, and the whence and whither of my journeying being duly performed, I changed my clothes for a miner's suit, and, lamp in hand, we proceeded to descend one of the several shafts by which the mine is worked. That "we," however, now included a Steiger, to whose care we had been confided. There are Steiger and Obersteiger, and (I believe) Untersteiger, their functions being to overlook the works and generally superintend the conduct of mining operations; their position varying between that of mining captains and of foremen or gangers. Though their pay is very scanty, averaging considerably under thirty shillings, and often not exceeding a pound a week, they have nearly all received an excellent technical training at the mining school, and possess an acquaintance with the theoretical principles of mining which it would be hard to find a parallel for among English miners of far greater pretensions. We spent some four or five hours underground, our conductor taking care that no instructive or interesting feature should be passed over, or be unappreciated for want of a commentary, and never tiring of explanations. The mine, of which the set contains five rich veins, produces zinc ores and pyrites, besides the argentiferous galena and silver ores, which are its main support. But though it employs over 1,000 men, it only turns out about 3,000 tons of ore a year, valued, according to the last return at hand, at some 45,000*l.* The sale of 80*l.* worth of " specimens " is one of the items which makes up this total. A generation ago, when only one-fifth of the present output was realised, it appears that the returns of ore sold were over 18,000*l.*, which indicated that the richest veins have been exhausted.

'As in most German mines, dead work bears here a much larger proportion to paying work than would be long tolerated by English adventurers. We find, by a recent return, that while only 1,000 metres were driven in the Freiberg Revier in rich ore ground, 1,800 were driven in poor though ore-carrying ground, and no less than 7,000 metres in perfectly barren ground. In other words, 70

per cent of the total year's work done was of an unremu-
nerative character. This mode of working, not for the
present alone, but with a view to maintaining the existence
of the mine for the longest possible period, has many and
solid advantages, which are not to be obtained on the
" quick return " system. Nothing gives a better idea of the
strong hold this desire for permanency has on those who
have the ultimate direction of mining works than the
extraordinary solidity and finish of the masonry which is
so largely used in the lining of the shafts, and the support
of the roof and sides of the working levels. The regular
thickness for the arches protecting the junction of galleries
with the shaft, or supporting the masonry of a few fathoms
of lined shaft, is one metre.

'It is the custom to inscribe the date on which any important sinking or driving was finished *in situ*, so that
the mine itself bears its own chronology graven on its
walls, and we have a clue to the exact course the works
have taken for a century or two. Thus, it will often
happen that at one stage in the descent of a shaft you
will find the date of say A.D. 1760; on getting still lower
you will be surprised to find you have got back to 1700,
and then, at the lowest depth of all, you are confronted
with a freshly carved or painted " 1876." This, of course,
indicates that in 1760 a shaft was sunk upon an old gallery
from another shaft (possibly only by accident, as it was not
continued down to the level), and that subsequently, the
original ore bodies being probably exhausted, the shaft has
been continued to its present depth, or a shaft driven
upwards.

'The shaft by which we descended was a rectangular
one, measuring two metres by six, and is to be carried
to a depth of some 500 metres. The main drawing and
pumping shaft, by which we ascended, was driven on the

veins, and follows its inclination, and is of very much larger dimensions. The greater part of the ore is got out by overhead stoping, though the underhand system is also in use. There is one tool which is very much used by the miners, which is not, I believe, common in England. It is almost exactly the shape of the ordinary miner's poltpick on a small scale (weighing only two or three pounds), and being held in position by the handle, is driven into the rock by a sledge; the handle enables the gad or wedge, which is what the tool really is, to be used in positions which it would be hard to get at otherwise.

' The Saxon mining lamp, though not unknown in England, seems such an obvious improvement on the naked candle, so largely used, that it is worth description. It consists of a flat box of wood, about eight or ten inches high, with a rounded top and the front open. The interior is lined with polished metal, and the open side may be closed with a glass sliding in a groove. This glass, when not in place, clips into a recess at the back of the lamp. Either a candle or oil-lamp can be used, and the whole is swung by a string round the miner's neck. The hands are left free, the flame protected from draughts and wet, and the light reflected on the work in hand. All these advantages are obtained at an insignificant cost.

' The miner's cap, common to all Germany, is of the shape once known in England as the " porkpie " hat, made of stiff felt, and is an admirable protection to the head, which, as every novice in mining knows, is exposed to grievous attacks in underground life. Gunpowder is alone used in blasting, and all the holes are put in by hand. As far as I could learn, Himmelfahrt is the only mine in the district in which machine drills had been fairly tried, nor do modern explosives seem much in favour. The miners are, by general testimony, as steady and industrious a

class of men as could be desired. Of late years Italian
(probably Piedmontese) hewers have been employed in the
Saxony collieries, and in driving adits and other heavy
work, and it is said that they can turn out more work than
the native miner. I was informed that a heading through
moderately hard rock, which we watched being driven,
was paid for by piecework at a rate which would give the
miner, a first-class workman, something less than 15s. per
week. The ordinary rate of payment appears to be a
mark (or shilling) for a six-hours' shift, and two marks for
a ten-hours' shift. Low as these wages are, they probably
do not represent a less purchasing power than the average
English mining wage. Indeed, they are even absolutely
but very little lower than the regular Cornish rates of a
few years ago.

'An excellent system of miners' unions, or friendly
societies, to which nearly all the men belong, contributes
largely to improve the position of their members. The
contributions of the men are supplemented by a propor-
tionate subscription from the various mining companies
and the income derived from various charitable endow-
ments. The distribution and management of the funds are
mainly undertaken, I was informed, by a committee of the
oldest members of the union. The objects on which they
are expended are : the relief, by allowances, pensions and
medical attendance, of sick members; pensions to widows
of deceased members; the maintenance of co-operative
stores, and the education of orphans and the children of
indigent members. The annual expenditure of the combined Saxon societies and foundations amounts to between
60,000l. and 70,000l. The whole body of ore-miners is
bound together by the Bergknappschaften, or unions—
which are of great antiquity—into a body corporate, with
elaborate regulations and ceremonies. To be expelled

from the association is the greatest social ignominy, and its established customs have the force almost of law. One of the periodical musters, or reviews of the Freiberg miners, was due a few days after I left Freiberg. On these occasions they are grouped into companies and brigades under their officers, adorned with the insignia of their craft, and, after attending church, spend the balance of the day in certain traditional exercises and festivities. Of late years a considerable tide of emigration of miners from Saxony to America has set in, and so relieved the pressure which the decrease of mining activity would have caused.

' *Saxon Mining*

' Neither women nor boys are employed in the metal mines of Saxony, and comparatively few in the coal districts. The Saxons, though rather a stolid race, are, as a rule, well educated, and believe in educating their children rather than sending them prematurely to work, a view in which the law supports them. The total number of miners employed in the ore mines is only about 8,000, but about twice that number are engaged in the bituminous collieries, and over 3,000 in the brown-coal mines. The colliers are a very different class of men to the ore-miners, whose *morale* and judiciously recognised *esprit de corps*, combined with a traditional good understanding with their employers, render labour troubles among them of very rare occurrence. I think there could hardly be a better indication of the old-world flavour which pervades Saxon ore-mining than the nomenclature of the mines themselves. A singular contrast to the matter-of-fact names which figure in our mining-share lists, and the ambitious and often grotesquely humorous labels which the Californian and Comstock miner delights in attaching to his workings, is afforded by a list, in which capital and dividends, and pro-fit and loss seem incongruous items, when connected with undertakings trading under such pious blazons as God's Blessing, God's Hope, Good God, Trust in the Lord, God with us, God trusted Daniel, the Green Twig and the Grace of God; sometimes lapsing into such mundane though comprehensive appellations as the Morning Star and Noonday Sun. Does not this seem to take us back to a far-off age, when work—or, perhaps, speculation—and religion were on intimate terms, though no one had yet formulated the " Gospel of Work "?

.   .   .   .   .   .

' *Saxon Metallurgy*

' The Fiscal Metallurgical Works of the Freiberg district consist of two great smelting establishments, one known as the Muldener Hütte and the once celebrated but now less important works at Halsbruck.   In connection with these there are certain subsidiary industries of considerable local importance, notably the Cobalt Blue Works at Oberschlema and Pfannenstiel (the latter of which is a semi-private undertaking).   The manufacture of shot and leadwork generally, of whitelead and pottery are the most flourishing of these subsidiary industries ; but they do not possess any features of special interest.   At the several Fiscal Works about 1,400 men are employed.   Tin-smelting is still carried on at six or seven small furnaces in close proximity to the mines, of which the most important are situated in the Altenberg district, but this branch of metallurgy is now labouring under considerable depression, owing to the fall in the value of tin. The Mulden and Halsbruck Works (which may be practically regarded as one), however, have certainly done more for the advancement of metallurgical science than any other establishment of the kind in the world, and possess many features of the greatest technical interest. The prominent position they have taken may be traced to a combination of several causes.

'In the first place, the intimate connection which has existed between the Academy and the Hütte since the foundation of the former, and the fact that for at least a century the direction of the works has been carried on under what, having reference to the current state of metallurgical knowledge, was unquestionably the best scientific advice, were alone sufficient to elevate the conduct of these works far above the dead level of empiricism which so long prevailed in metallurgy. The joint reputation of the Academy and the Works also brought to Freiberg a constant succession of intelligent visitors, whose suggestions for modifications of any process or accounts of the modes adopted for like ends in other countries were always attentively considered by experts, whom an academy training had freed from local prejudice, which so often prevents the adoption of improvements. The remarkable complexity of composition, which is a characteristic of the Freiberg ores, also calls for the exercise of an unusual amount of skill in devising processes by which the largest number of metals may be profitably isolated from each other and turned out in a marketable condition. The absence of those restraints upon the pursuit of investigations of which the immediate pecuniary result is doubtful, more or less inseparable from private enterprise, has also had a most happy effect on Saxon metallurgy.

' During the most prosperous period of the Saxon mines the ores were smelted at a number of private works in a very rude fashion. Towards the commencement of the eighteenth century, when the succession of rich bonanzas which had astonished Europe and enriched Saxony had been about worked out, and the effects of the vast importa-

tion of silver from Mexico and Peru in depreciating the
value of the metal had not been recovered from, the Saxon
Government came to the rescue of the impoverished mining
industry by founding metallurgical works, under the
administration of a special department, with the object
of utilising to the utmost the mineral treasures of the
Erzgebirge, by bringing the advantages of capital, concen-
tration and skilled management to bear upon the extraction
of the metals from their containing ores. The results of this
direct Government interference with private enterprise,
repugnant as it is to English ideas of the limits of the
functions of the State, have been certainly more favourable
than could have been anticipated. Aided by the economical
results achieved by the Government works, of which the
miner shares the advantage, not only in receiving originally
a better price for his ore than private smelters would or
could give, but by a subsequent participation in the profits
of the undertaking, many mines have struggled through
periods of adversity to which they must have otherwise
succumbed.   In looking over the visitors' book at the
Muldener Hütte, one is struck by the cosmopolitan
character of those who (as indicated by their names) avail
themselves of the unreserved liberality with which the
direction permits access to all the Government establishments. My own visit was paid in company of two Greeks,
our predecessors being Germans, Spaniards and Americans.

.   .   .   .   .   .

'Of the 130 ironworks of Saxony—of which only some
half-dozen have blast furnaces—located for the most part
in the neighbourhood of Zwickau, Chemnitz, and Plauen,
with a gross production valued at about one million
sterling, I have no personal knowledge. I was informed,
however, from several sources, that, notwithstanding
journalistic denials, the engine and machine makers of

Chemnitz and Leipzig always use English metal, especially steel, for any purpose in which the highest quality is required.

'From Freiberg to Dresden the railway passes through decidedly attractive scenery, while, for the technical tourist, the attractions of the picturesque valley which the line traverses are not diminished by its being the seat of a thriving brown-coal mining and iron-working industry at Potschappel, and the celebrated forest nursery and forestry academy of Tharandt.   On the many attractions of Dresden, the most charming of German cities, this is not the place to expatiate. It may be suggested, however, that the geological and mineralogical collections which form, perhaps, the least frequented section of the magnificent series of museums of which the Saxons are justly proud, are worthy of their reputation, their strength lying in the completeness of their sets of Saxon ores and fossils. The Saxon Switzerland, which commences a few miles south of Dresden, originally an elevated tableland of sandstone, has been chiselled, by fluvial and aerial agency, into a series of fantastically-shaped peaks and pinnacles, and isolated and precipitous rock fortresses, while those portions which have suffered least are penetrated in every direction by deep ravines. As it is given only to few to visit the Colorado canyons, an excursion to the Sächsische Schweiz may be taken to be, perhaps, the most favourable accessible illustration, on a great scale, of the power of water as a geological tool, since the cause and effect are here seen in close juxtaposition, and under the most striking conditions.

'Bohemia, a country which lies somewhat out of the regular tourist track, holds out many inducements to the student of metal'urgy or mining who has got so far as Freiberg or Dresden to extend his explorations thither.

Amid scenery often in the highest degree wild and picturesque, mining has been carried on in Bohemia for considerably over a thousand years.

'In the narrow gorge of Joachimthal, where the first thalers were coined, and whence their name is derived, may be seen mines still in active work, producing silver, lead, cobalt, bismuth and uranium, in which some forty successive generations of miners have laboured. Near the fine old city of Prague, one of the most interesting in Germany, are the wonderfully rich silver-lead deposits of Przibram, which have been worked continuously for eleven centuries. Large deposits of lead, and smaller ones of copper, tin, and cobalt, are also mined in many other districts of Bohemia, the systems of exploitation and dressings being hardly, if at all, inferior to those adopted in Saxony and the Hartz. Indeed, much of the most approved modern dressing machinery has its origin in Bohemia and Schemnitz.

'The iron industry of Bohemia is of hardly less antiquity than its silver mining. Great deposits of hæmatite and other iron ores are spread over the country, the ore being smelted chiefly in charcoal furnaces close to where it is raised. In no district in Europe is the charcoal blast-furnace seen to greater advantage than in Bohemia and the adjacent Austrian States. At Kladno, however, and elsewhere, coke furnaces have been recently erected on a considerable scale. Though both bituminous and anthracite coal is worked to some extent, the chief fuel resources of Bohemia are found in the enormous supply of brown coal which it possesses, much of it consisting of deposits considerably exceeding ten yards in thickness. It is now about eleven centuries since the Bohemian gold-fever broke out, and the washing and digging of that day appear to have been pretty thorough, since nothing

has been left for their successors but heaps of washed sand
and gravel. In short, the metalliferous industries of
Bohemia are hardly less varied and interesting than those
of Saxony ; while by extending one's excursion to Hungary
on the one side and Styria and Illyria on the other, one
would have a tour in which an absolutely complete acquaintance with all that is remarkable in Continental mining and
metallurgy might be gained, in conjunction with an ex-
ploration of the almost unique beauties of the Austrian
Alps and the Hungarian forests and highlands.

'But there is another region of Germany, very much
more accessible from England, almost, indeed, at our
doors, which possesses within a very limited area many
very diverse claims on the attention of the sober holiday-
seeker. The Hartz offer a rich harvest to the geologist,
mineralogist, metallurgist and miner, and have no mean
attractions for the artist and antiquarian. Till some
twenty years ago a region almost entirely primitive and
out of the world—the summer hordes of Berliners, Ham-
burgers, and other denizens of the plain, who have since
been induced by railway facilities to invade its more
accessible districts, have not yet succeeded in changing
entirely its former character, though the simplicity of the
inhabitants and quaint picturesqueness of its towns will
probably soon be things of the past.

'Eisleben, of which the principal claims to distinction
are that it is the birthplace of Luther, and the seat of
administration of the Mansfeld'sche Kupferschieferbauende
Gewerkschaft, fairly illustrates the close juxtaposition of
things new and old, so apparent to a traveller in the byways
of Germany. In the architecture of the town, the Luther
period is the most prominent; in its life, nineteenth-
century industrialism. The Mansfeld Copper Company,
which now carries on the mining and smelting of the

copper schists, which were first attacked in Hesse in the
tenth century, and at Eisleben in the sixteenth century,
by the Counts of Mansfeld, is a consolidation of five
companies, united under one management some five-and-
twenty years ago, which now, under the direction of
Bergrath Leuschner, has the reputation of being one
of the best managed, as it is one of the most prosperous,
industrial corporations in Germany. In 1876 the company managed to earn the very respectable sum of
126,000*l*., giving a dividend of 37*s*. on each of the 69,120
shares into which it is divided.

'Over a considerable area of Central Germany there is
found a fossiliferous and bituminous marl-slate, covered
by the *Zechstein* or magnesian limestone, and overlying
first the *Weissliegendes*, a sandstone containing in places
small quantities of copper, and under this again the
Rothliegendes, a red sandstone mixed with conglomerate,
basalt, &c. These deposits lie in a great basin, and at
various points on the rim, where the marl-slate crops out,
attempts have been made to work it for the copper which
it contains, mainly as pyrites. It is only, however, in the
neighbourhood of Mansfeld and Eisleben, where an undu-
lation in the strata brings a large quantity of this slate
within a short distance of the surface, the dip being only
about 6°, that it has proved permanently to pay for
extraction. Indeed, even here it is only by working on
the largest scale—the Mansfeld Company raising last year
235,000 tons of cupriferous schist and sandstone—by which
the standing charges are spread over an enormous output,
that remunerative results are obtained.

.   .   .   .   .   .

'The works and mines together give employment to
8,000 men. The system by which this army of *employés*
and their families is supplied with the necessaries of life

G

by the company is well worthy of attention. Throughout
the Hartz district the mine-owner, who is for the most
part the Government itself, is looked to to supply the
necessaries of life, or at least the chief of them, to those
he employs. The reason of this custom, which has prevailed for centuries, is to be found in the fact that the
forest-clad hills and bleak tablelands of the country are
scarcely capable of bearing corn enough to supply the
wants of the sparse population which cultivates them,
leaving no surplus for the mining population and its
tributary industrials. Thus, imports of grain on a large
scale have always been necessary. So we find the Mansfeld Company distributing annually nearly 4,000 tons of
rye-meal to its workpeople, or at the rate of over a
hundredweight per man per month. Rye-meal at Mansfeld costs nearly 9l a ton. It does not appear that this
peculiar modification of the " truck " system, by which the
employer undertakes the duty of feeding his men as well
as paying them wages, has been accompanied by any of
the abuses which seem inseparable from it in England.

'A benevolent, or friendly society, not less admirable in
its provisions than that which exists at Freiberg, is in active
operation here also. To its funds the company contributes
largely, no less a sum than 8,000l. a year being at present
devoted to this purpose, besides a considerable sum spent
in special gratuities and allowances in cases outside the
regular operations of the society. The amount of the in-
vested funds of the society at the beginning of 1877 reached
the satisfactory sum of 27,000l., while the disbursements
during the year 1876, in pensions, sick-pay, medical relief,
&c., amounted to over 16,000l. · Thrift is fostered by a
savings bank, in which the men are encouraged to deposit.
It appears, however, that only some 800 of the 8,000
employed are depositors, the average deposit being about 6l.

' From whatsoever point of view it is regarded, the Mansfeld Copper Company may fairly be considered one of the most interesting of the great industrial establishments of the Continent. Having successfully solved, thanks to the persevering and unassisted investigations of its own officers, some of the most difficult problems of metallurgy, no one can deny that it deserves to enjoy the prosperity to which it has attained, while its management continues to be marked by the same technical skill and energy, and care for the welfare of the employed, which now characterises it.

' From Mansfeld it is a four hours' walk, through a not very attractive region, to Harzgerode, where the beauties of the Hartz really begin. In the vicinity are several silver-lead mines, which changed hands at high prices during the company mania which raged so fiercely after the war, but have not proved much of an acquisition to the Berliners into whose hands they finally passed. A beautiful walk through a hilly and richly wooded country brings one to the old established ironworks of Madesprung; and after traversing a long stretch of closely wooded hills, we arrive at the flourishing little town of Thale. Thale occupies a very advantageous position on the extreme border of the great plain which stretches away to Berlin and Hamburg, at the point where the river Bode emerges from the wild and singularly picturesque gorge which it has cut through the mountains, which at this point rise almost perpendicularly from the plain. It is the terminus of a railway which brings every summer a yearly increasing crowd of visitors, attracted by the beauties of the Rosstrappe and Bodenthal, and which by placing it in direct communication with Hamburg, Magdeburg, Berlin, and the Prussian coalfields, puts this little town in a position to develop the industrial position to which it has

G 2

already begun to aspire. An abundance of water from streams which by a slight diversion of their course might be made to yield considerably more water-power than is at present utilised; enormous supplies of wood and charcoal from the adjacent hills, which also contain large deposits of iron ore; these, with cheap labour and comparatively cheap land, make Thale a place worthy the attention of manufacturers.

'Last summer the Thale ironworks, which are in the hands of a company, were in brisk work, turning out bar iron and rods, light rails and plates, and, I fancy, wire, and a large variety of small forgings. A small establishment adjacent to the ironworks, occupying itself apparently chiefly with agricultural implements, and remarkably well supplied for so small a place with machine tools, was also well occupied, being engaged in turning out in large numbers a very convenient kind of light iron wheelbarrow of very convenient shape and easy to handle.

'A mile or two on the road to Blankenberg I found a small brown coal pit being vigorously worked; a powerful portable engine was engaged in hauling the trucks of coal up an incline and at the same time driving a centrifugal pump by which the pit, which was an open working, was drained. Indications are not wanting of the presence of a brown coal not many degrees removed from peat, in many localities hereabouts, and if worked in the inexpensive but effective fashion I saw in operation it must be a cheap and useful source of fuel.

'Blankenberg, a quaint old town with steep streets and a picturesquely-dominating chateau, is another border-town of the Hartz which is being rapidly invaded by the new ideas that follow in the wake of railways. Some three or four miles from the town, among the hills, are great beds of ironstone, in a situation almost inaccessible from the

steepness of the roads leading to them. By means, however, of a tramway carried through the hill by an expensive
tunnel, these have been reached, and two first-class blast-
furnaces erected on the edge of the plain to melt the ores
raised from them. Projected during the epoch of inflated
prices and feverish prosperity in the iron trade, it seems
that these furnaces have had a hard struggle to secure even
an entry into the arena of competition. Last summer
there was every indication of a shortness of funds having
been encountered even before their completion. As there
was, at the time of my visit, no one on the works in a
position to give any reliable information, I could only get
a general impression of the intended arrangement of the
furnaces. The furnaces appeared to be designed as cupolas
of good modern design, with four tuyeres, a. slag-hearth
at the back, a water balance hoist, a central gas-tube, and
excellent blast-stoves. The blowing engines, of the hori-
zontal type so popular on the Continent, are particularly
fine ones, and there is abundance of room for dumping the
ore, which appears to be of excellent quality, storing coke,
and forming slag tips. A branch railway has been constructed to the furnaces, by which they will receive fuel
and send away their iron.

'Leaving behind this infant establishment, designed
on the most modern and approved principles, and
representing an enormous expenditure of money, but
having, it is probable, far from bright prospects of success,
it was curious to find in the midst of the hills, not many
miles away, another ironworks, ancient, primitive, with
no expensive plant or modern facilities for carriage,
and yet busily occupied and flourishing exceedingly.
The Rübeland Hütte, in a beautiful situation in the
valley of the Bode, almost confines itself to the manufacture
of castings, for which it has a great reputation. The ore,

partly hæmatite and partly brown ore, containing from 30
to 40 per cent. of metal, is brought in carts from workings in the vicinity, and smelted in low and old-fashioned
blast furnaces, of which one is now worked with coke, the
other with charcoal. The blast cylinder, a very ancient-
looking machine, is worked by a water-wheel, though this
sometimes fails in dry summers and severe winters. The
charcoal, of which large quantities are used, is made in
iron retorts, the tar and other products of distillation
being collected and sold. This mode of preparation is
found considerably more economical than the ordinary
system of burning the wood in heaps. I was informed
that an average yield of twenty to twenty-five per cent. of
charcoal is obtained in the retorts, against only fifteen or
sixteen per cent. in the meiler, but this latter yield seems
unusually low. The manager, a Freiberg graduate, stated
that it required something over twenty hundred-weight of
charcoal to produce a ton of pig-iron; with good blast-
stoves and improved furnaces, probably a fourth of this
consumption might be saved.

'There is an enormous demand throughout Germany
for cast-iron stoves, and the Rübeland Foundry is largely
occupied in supplying these. The design of the ornamental
open-work castings of which the sides and fronts of these
stoves are constructed, offers a good opportunity for the
exhibition of taste and skill, and some of those I saw in
the storehouse were really fine specimens of art workman-
ship, and the perfection to which castings in iron (which
is, perhaps, of all metals the most suitable for taking
accurate reproductions of intricate patterns) may be
carried. The moulds are made in a material which seems
intermediate between our own loam and the celebrated
casting sand used in Berlin. Some of the castings are
made with the metal run direct from the blast-furnace,

others after remelting in cupolas in the ordinary way. The
ores here contain a considerable amount of phosphorus,
which may probably contribute to render the iron suitable
for fine castings.

    •   •   •   •   •   •   •

    ' Clausthal, now the most busy of the seven mining
towns of the Hartz, having in its recent technical activity
far outstripped the venerable imperial city of Goslar,
possesses no ordinary interest for the student of mining
science and advocate of organised technical education.
The Mining Academy, with its museum, the Aufbereitungs-
Werke, or dressing-floors, the mines and their drainage
adits, and finally, the smelting works, are each among the
most instructive of their kind. Of these various institutions the Mining Academy is perhaps the most worthy
study, as offering an example of what such an establishment should be, not less instructive than that of its more
celebrated rival at Freiberg.'

    The articles close with an elaborate comparison between
German, Belgian, French, and English metallurgical
schools.

CHAPTER VIII

EXPERIMENTS—A DASH INTO SWITZERLAND

UPON his return from Germany, Thomas again pressed Mr. Gilchrist to undertake experiments. A little later in the year he spent a few days of his remaining leave at Bradford (in view of the autumnal meeting of the Iron and Steel Institute in Leeds) ; there he met Mr. Gilchrist. The projected experiments are spoken of in the following letter.

'Thames Police Court, 1876.

'Dear Bess,—Last week I had five days at Bradford, which I found a pleasant break. The Iron and Steel Institute were holding their meeting at Leeds, and I went over every day nearly. One day a picnic at Kirkstall Abbey, and so on, the ironmasters of the neighbourhood coming out strong in hospitality. . . . Percy also at Bradford for the meeting. . . . I go down to him for a few days if I can get away, to try some experiments which are at present engrossing all my attention.

'I have just finished some rather elaborate technical articles for "Iron," and am going to take a rest. . . . Yours,

'S. G. T.'

During this autumn, Mr. Gilchrist left the Cwm Avon Works and removed, still as analytical chemist, to the Blaenavon Works, then under the management of Mr. Edward Martin, who was afterwards to play a considerable part in the development of the basic process. Curiously enough, Thomas was a friendly competitor with his cousin for the Blaenavon appointment. Mr. Martin selected Gilchrist because he was a 'practical' chemist, and Thomas apparently was not. In the preceding July, Sidney had failed to be elected a 'Fellow of the Chemical Society' [1] on a similar ground, because he declined to describe himself as a chemist, when he was a police-court clerk.

On December 20, 1876, Thomas writes to his Blaenavon cousin, making certain financial proposals and saying:—

'I have not been able to make any head with private steel-making. I still cling to the idea that *our idea* has something auriferous about it. . . . Whether we shall either of us be able to devote the time to it it requires (and I find the coin) is quite another matter. I am always expecting some wretch to walk in and do the thing.'

Mr. Gilchrist answered on the following day :—

'My dear Sid,—I think your proposal too advantageous to me. I really hope in January to manage some experiments with it.—Yours,

'P. C. G.'

This Christmastide, Thomas writes to Wiesbaden in a somewhat despondent tone, perhaps because so little progress has been made during the year with the enterprise he had so much at heart :—

To Miss Burton

'18 Sussex Place, December 1876.

'Dear Bess,—All good wishes for '77, and all thanks for your good wishes for me. I can't say I have any very

---

[1] He was duly elected in June of the following year (1877).

brilliant anticipations for my own, short of the achievements of the year, which I regard mainly as a bore succeeding to another bore. It is pleasant, however, for once to know of your so enjoying yourself at the festive season. We have had it rain continuously here for the past month, a state of things which, though gloriously grumbled at, doesn't seem to me undesirable.

'Went yesterday to inspect a real ice rink, established in a floating structure on the Thames. Had a copious interview with the inventor, who seduced me into an experimental tour on skates. The place thronged (only holds thirty or forty) four times daily for two days a week at five shillings per two hours. The apparatus by which it is arranged, very interesting. . . . I have just finished "Our Mutual Friend," which I have protracted over a period of three weeks as a prandial *bonne bouche* with immense enjoyment. I meant to send you the annual by Farjeon, who is a colonist from New Zealand who aspires to be another Dickens. These tales, however, made such a hit, that every copy was sold before I could secure one. I have a short note on Freiberg this week, which I will send you. Llewe has now published a pamphlet, a very good one, which I as critic duly "noticed." Such is life.—Yours,

'S. G. T.'

The new year of 1877 crept on with little done for dephosphorisation; but in the early summer of that year, Mr. Gilchrist began experiments in good earnest, Thomas constantly (as his letters show) criticising results from London, and suggesting further trials.

The following epistles to Wiesbaden belong to the earlier part of 1877 :—

To Miss Burton

'Dear Bess,—Lil was immensely pleased with your music. . . . I should like to hear you again. I heard scarcely anything from your gorgeous ebony instrument. We shall be flitting certainly in June; so, unless you are speedy, you will never see us in our West End mansion, but rather in some tiny domicile in the most unfashionable of quarters. . . .

'Have been reading Browning, so feel more than usual difficulty in writing anything intelligible. Met several Australians at G——s the other night. They are fervent in praise of the antipodes, so we got on well. . . .—Yours,

'S. G. T.'

'18 Sussex Place, Onslow Square, London, S.W. 1877.

'Dear Bess,—We have grown bad correspondents; you, I am inclined to think, being considerably the worst, though you have fresh excuses to tell of and I only old ones. So you won't pay us a visit this summer? Oh that furniture mania which obstructs so much that is desirable! When I establish a house of my own (in the Far West, Australia, or Africa) my chattels will certainly be confined to a fold-up campstool and possibly a portable table and a tin can. I loathe town more year by year. My colleague proposes to settle some twenty-five miles down in Kent. A letter this morning from my ex-professor in America, now "Metallurgical Manager" in Colorado, urging me as usual to go out and make my fortune. . . .

'The lady medical students in London have gained their long desired objects—a hospital to study at and a right to enter for the two degrees, the London University M.D., and the Dublin Medical degree. I have just finished Bulwer's "Parisians," which I am inclined to

believe is his best novel—though his political sentiments
are very far from being mine.

'I am not very brilliant in a sanitary point of view;
talk of running down to Wales for a week, if I can get
away in May or the end of this month.

'I have been doing little in the scribbling business
but pure and bristling technicality, and of that I am
pretty tired. Miss Martineau's life is chiefly autobio-
graphic; it has caused some excitement. Her criticisms
are anything but flattering on her distinguished acquaintance; bishops, lords, lawyers, and authors are impartially
dissected. I have been reading also a curious book on
Spain, which makes one think Spain a country worth ex-
ploring. Lily is making me groan under the burden of
social duties; has absolutely led me into two dinner parties
lately. I hear A. H. thinks Wiesbaden Elysian.—Yours
ever,

'S. G. T.'

'18 Sussex Place, May 25, 1877.

'Dear Bess,—I have been house-hunting *ad nauseam*,
productive of nothing but weariness and disgust. I have
found several which would suit according to my modest
views; but the M. and Lil are not so easily satisfied.

'Lil went to the Hospital Ball last night. . . . The
G——s chaperoned her. I cried off, the effort being too
much for my endurance to be bored for six consecutive
hours. . . . Calling on a man last night, I was dragged
off to a *Bradlaugh* meeting, that very vigorous contro-
versialist having been persecuted for the publication of a
rather incisive and vigorous pamphlet on an important
socio-physiological topic. I anticipated being bored, but
found it great fun. Bradlaugh an orator, I find, of
singular readiness and force. Several ladies who have

espoused his cause spoke admirably, and the proceedings were enlivened by some students—medical—making a disturbance resulting in a fight and general *mêlée*.

'Other news comes but slowly, and events seem to drag. MacMahon in France has blundered to an extent which must be satisfactory to your German friends, and will probably on the whole duly serve to consolidate the Republic and the anti-clerical party.

'I am going down to Sevenoaks on Sunday to see a place my colleague has taken there, and which he vaunts as a very paradise.

'Wiesbaden will be looking just charming now before the baking season has set in; not so London.

'I am not defined on my holiday plans— shall probably go to France or stop in England. Have had no time to touch German since last summer, and have forgotten the modicum which then served me. Have been very seedy indeed for some months; had to vegetate under medical threats of dire pains and penalties.—Yours,

'SIDNEY G. THOMAS.'

Here again we have the warning note presaging the ultimate breakdown in health.

In June 1877 Thomas, as already noted, was elected a Fellow of the Chemical Society. In this month, too, the household removed from Sussex Place to Queen's Road, Battersea, where was the family dwelling-place for some two or three years to come. Shortly after this removal Thomas went abroad for his summer holiday, to be spent this time among the Belgian ironworks, with the late Mr. Edward Grosvenor as a companion—Sidney being armed by Mr. Chaloner with a letter of introduction accrediting him as the authorised correspondent of 'Iron.' He was unable, however, to resist the temptation of a preliminary

dash into Switzerland on his own account, and the following
spirited letters therefrom may come pleasantly to some
readers.

To his Mother [2]

'Meiningen, 1877.

'Dearest M.,—You see the mountains prevailed, and
here am I, finding Schweitz even more unique and lovely
than I had imagined—far before the Tyrol. I have been saying all day what a shame I should be here and the M. and
L. at Battersea. Left Wiesbaden on Sunday at noon, sur-
feited with hospitality almost. Had a heart-rending eight
hours' ride to Strasbourg. Walked about the town &c.
till 2 A.M., then to Basle; on again to Lucerne, which
looked just charming, like the drop-scene in the opera
(music and all). Then a delicious sail up the lake: each
turn fresh sets of beauties. Landed at Alpacht. By
coach to Lungern (this coach a concession to you, of which
I was thoroughly ashamed). A ruinous and gorgeous
dinner (4s.) and then walked here, picking up a Scot on my
way—then a Swiss, with whom I am now on intimate
terms, if I understand him rightly, sharing his room &c.
Our window looks on superb waterfalls and the snow-clad
Wetterhorn. Write to Chamounix.'

To his Mother [2]

'Niederwald, 1877.

'Dearest M.,—I fear you will have grumbled at pencil
scrawl, but ink was at the moment unprocurable. From
Meiningen (my last night's quarters), I started at 3 A.M.
with my Swiss, soon picking up a *Fahrer* as a companion
(not as a guide). When my Swiss spoke before it made
my hair rise to understand half his speeches; but when he
got talking to the *Fahrer*, he became a linguistic sphynx.

[2] Written on a post-card.

With rests and coffee on the road we passed over the Grimsel
(near 7,000 feet high) and had investigated the great snow
glacier by 1.30 P.M.   It was a respectable walk and climb,
two hours being in the 'tarnal' snow, which nearly
blinded me with its glare.   The scenery a succession of magnificent pictures, glaciers, wild rocks, torrents, waterfalls
(of a size and beauty to make the fortune of an English
county).   The hospice not far from the top, with 4-feet
walls, where two nuns exist all winter through as receivers
of the lost, dogs, &c., in orthodox style,—we using it for
coffee supplies only.   At Rhone glacier I adieu'd my Swiss,
as the *Fahrer* was becoming a bore, and took a long piece
summâ diligentiâ, which is an excellent way of seeing the
country, though extravagant.   I turned in here to
country inn (not hotel), and have just discussed four eggs,
salad, wine, cheese, &c.   The room, with walls and ceilings
of painted wood, has long windows from which I see first
a great stretch of green slopes (the infant Rhone intervening with turbulent roar), the pastures dotted with
chalets, magnified copies of those you have: then, higher,
a fir-wood: higher still, rocks and great patches of snow,
a few waterfalls thrown in.   Roses outside the window
and in the room.   Would you not enjoy it, and Lil, and
A. ?

To his Mother [3]

'Martigny, 1877.

'To resume.   At 5 A.M. started for Viesch, from whence
a long pull up to a hotel, some 6,000 feet.   Three young
Englishmen outside, more inside: in fact all English.
Then to a glacier, when a climb !   On way, met a girl and
her father, who thus from a distance :—" I suppose you
speak English : if so, don't go that way."   However, I did,
and got rather in a fix, but extracted myself, and getting

[3] On a post-card.

to top, had a glorious view over the greatest glacier in
Europe, a lake of ice, and some score of snow peaks.
Then down. Of course I would not stop at the hotel with
English mob; so, after copious milk at cowherd's chalet,
I adjourned for the night to a hay-chalet, where I saw the
sunset to perfection, and rose from my hay to see it rise.
My first camp-in (or out) a great success. This morning,
down to Viesch, and *Frühstück* at a pleasant new hotel,
where alas, a maiden who to a rather nice face added perfect English (gained as nursery governess in Lancaster).
With whom a long chat, followed by a heavy disbursement
(comparatively). Tearing myself away,—by the Rhone
—back to Mörel. The Rhone rapid and turbulent, between rocky banks, and the high valley sides forest-clad on
each side. Most interesting; though I confess to being
haunted by the Yankee idea of utilising its fierce currents.
Horse's and man's muscles should alike be spared here.
Here, near the entrance of the Simplon, German, French,
and Italian meet. The climate Italian. Grapes and
chestnuts &c. make the valley greenest after a five-course
dinner and a pint of wine. I wonder what the *rechnung*
will be? Have just been out-chatting to the passengers
of the passing diligence. All English. Would you could be
here. Have been discussing with Italian metallurgists
Italian metallurgy. Our views differ.

To his Mother

'Au Touriste, Les Figues : Saturday.

'Dearest,—At an open window, looking over a small
wood direct on to the Mer de Glace, which is backed up
by the Hignelle, sharp pointed rock, 10 and 13,000 feet
high; the side window, also my bedroom window *vis à vois
de Mont Blanc* (as my landlady says). Once more, here
is a place where you ought to be. To resume my postcard

diary. Just as I finished my card to you on Thursday, two young Scotchmen in regular tourist style came into the 40 'salle à manger,' of which I had before been sole possessor. We struck up an acquaintance at once— gentlemanly fellows from Edinboro', law-students I fancy. Had a lot of tourist talk and great fun over ordering their supper and a bath for next morning. I found they only mustered about thirty words of bad German between them; so I, with my sixty, came in as a swell linguist and deliverer. The bath floored us all. However, the girl knew a bath, such as they have establishments of, and she knew the slop-basin, which is the regular substitute for a basin. I explained (or thought I did) a wash-tub would do; she then would have it we wanted a saucepan, and so on, till I laughed my viscera into jelly. Next morning up at 4.30, couldn't get breakfast till 5.30, so felt awfully late; hadn't gone to bed till 9.15, which also made me feel dissipated. I and my Scotchmen parted, they to do the Eggischhorn; they had been out three weeks, had three weeks more. I told you about my charming waitress and ex-governess. I implored the Athenians to call on her and part freely with their bawbees for the good of the house! After a smart run of five miles caught the dili- gence as it was leaving Brigue, and found myself suddenly among French-speaking folk, or at least folk who speak French first, and German and Italian with equal ease. I found I could not muster ten words of French—kept re- lapsing into German, and then making a hash of both till I bewildered the conductor to perfection. A long drive along the Rhone valley; here flat and marshy, though with big hills on each side, little clusters of chalets perched up in places where you would think everyone must be always giddy and hold on by the grass. Fearfully hot; turned out at a new railway-station, and on by rail

H

through more steaming valley, with ruins now and again
along the hillside, till we got to Saxon, where I turned out
and wandered in full marching costume into the Casino,
where some two hundred well-dressed people, mostly
middle-aged and oldish men, and some twenty middle-aged
and young women, with diamonds &c., were hard and deep
at *Rouge et Noir* and *Trente et Quarante*, earnest and intent,
and calculating as if their lives depended on it, the women
only going through the routine of smiling when they lost.
It was a sight not to be missed. The croupiers excited my
admiration for their quick eyes and calculating powers.
After an hour thus spent (and *without* staking the regular
five francs) I moved on, my movements (a sort of Robinson
Crusoe in a ball-room) being quite attentively watched
and commented on. A hot five miles to Martigny, where
(at entrance to St. Bernard Pass and that to Chamounix)
I moved on some 3,000 feet up to a tiny restaurant
where I found two French families (eleven persons)
en pension, and yet a diminutive room for me. The
French families very polite, painfully so, inasmuch as I
found I could not put two words together without German
interpolations; the terms four francs a day. They seemed
wonderfully happy, Papa telling me that with a *glacier*,
*les bois, les montagnes, et les voyageurs passants, les vaches
et les chevaux*, what could children want more? To which
I replied, *Pas de tout.* I, however, got charged six francs,
and didn't get a dinner. (N.B.—I had had a big dinner
at the station. N.B. N.B.—I am feeding prodigiously;
if I did not walk it off I should speedily emulate Daniel
Lambert.)

'This morning soon after five of the eleven had tortured
me with *Bon jour, monsieur, j espère que vous avez bien
dormi*, I bolted from the salutations of the other six, and
trotted down into a valley, and then up another 7,000 feet

high pass, the *Col de Balme*, where I invaded a dirty
hovel in which butter and cheese were in process of manufacture, and consumed about two quarts of milk, to the
astonishment of the very grimy proprietor. Then down
the other side and investigated a glacier which possessed
a good big waterfall and a *moraine* (vide Arthur) which
evolved, when I determined on examining it, about ten
feet in height and fifty long; but I found it took me
half an hour to climb to the top. Then on down the
Chamounix valley to this place, a roadside inn, which I at
once perceived would suit my purse and tastes better than
Chamounix. Old-fashioned people and place. Have just
had a monstrous *café complet* (*i.e.* about a pint of milk
and coffee, bread, butter, and honey), to which I added
five eggs! Call no man happy till he dies or I should say
I'd made a discovery. Switzerland (by the way this is
France) might be called Cow Land, cows and travellers
being the staple industries. The cow-bell is everywhere,
at the top of the hills and the bottom of the vales, ever
tinkling, not unmelodiously. On the hills a man or boy
has charge of some twenty cows, by the roadside a boy or
girl has one or two. The cow-girls knit by the way
generally, and have an eye to passing business. Thus to
me, one: " *Bon jour, monsieur; monsieur est fatigué, n'est-
ce-pas?* " I : " *Non, merci, pas de tout* " (N.B. That was a lie).
" *Ah, non ? J'en suis heureux* (there's sympathy for you) *car*
—meditatively—*si monsieur veut prendre quelque chose*—
comme (piano) *un* (diminuendo) *petit verre, mais bon . . .
Monsieur sait qu'il y a une auberge avec de bons lits près
d'ici.*" . . . I involuntarily exclaim, " Ah ! " " *Et je vais y
conduire monsieur,*" which she (and the cow) proceeded to
do. Five young women have just passed, separately,
taking four individual cows to their slumbers; one cow
had two guides, one holding her tail (the cow's) and

H 2

knitting, and one holding her horn and ditto. The *Mer de Glace* has been going off, and cracking considerably. It is a curious noise, like musketry fire. By the way, mindful of my promise, I am doing no hills, or anything else with a tenth per cent. of a spice of danger, which is painful but meritorious. Poste Restante, Liège, Belgium, my next address. I hope to get at least one letter from you to-morrow, and to hear you are blooming. Your letter of Saturday I've just had. I am miserable; it is eleven; I've been up since five, and it has been pouring all the time. I shall have to go to church shortly! It is dreadful! The place steams!—Yours.'

To his Mother [4]

'Chamounix, Sunday, 5 P.M.

'Desolation! Misery! *Toujours la pluie.* I went to church: first looked in at Catholic, but found them steaming full; then at English, a rather pretty building where I found some 120 of my compatriots, dressed *à outrance* and going right through the whole service as though they had been in a Queen's Gate Tabernacle; two clergy, conventional sermon, piety rampant. For myself my leggings, alpenstock, waterproof and pockets stuffed with books and papers, constituted an individuality. Got this afternoon your card as well as letter. So pleased all is going well. I have told them to send on any other letters. Just had an excruciating conversation with hostess. I am rapidly aging under these efforts. I confirm her idea that we do not see the sun for nine months, chiefly because it is easier to say *oui*! She informs me meat is dreadfully dear—9*d.* a pound. Oh, the misery you caused by abstracting my old leggings; there is a void of three inches

---

[4] On a post-card.

which makes me vulnerable, which they were destined to cover. Let it be a warning!

'The *Mer de Glace* looks as if the rain did not agree with it any more than with me. Shall be at Liège on 20th and 21st. I am thinking how I could run a railway up Mont Blanc, and work it by the stream at the foot. The superfluous water power here torments me.

'*Tuesday.*—Did *Mer de Glace* on Sunday after all.'

The visit to Belgium was as pleasant to Thomas (although in different fashion) as that to Switzerland. Mr. Grosvenor speaks of the delightful enthusiasm with which Sidney explained to him the working of the Cockerill manufactory at Seraing, where the travellers were—thanks to Mr. Chaloner's letters of recommendation—received by M. Greiner with great hospitality.

CHAPTER IX

THE BASIC PROCESS PUBLICLY ANNOUNCED

ON our inventor's return to London, we find him again in constant communication with Mr. Gilchrist.   On September 11, 1877, he writes :—

' Have some idea of going to Newcastle, just for a change.   Have been uncommon seedy for past fortnight ; have just struggled through work at Court, that's all. Sore throats and so on are making life a misery.   " P " is a great and promising subject.'

On October 2 he writes again :—

' I fear question of blast will be troublesome.   I made a lot [of] inquiries about blowers.   How would the steam engine answer by reversing its action ?   Don't laugh. Instead of the steam driving the piston, would not the blast be turned on instead of steam ? '

Thomas did escape to Newcastle, to the autumnal meeting there of the Iron and Steel Institute, as here projected.   Mr. Chaloner was with him upon this occasion. He well remembers Sidney's going, during this expedition, to the theatre at Middlesbrough, and being much affected by Miss Jennie Lee's wonderful impersonation of ' Jo '— an impersonation which has moved many men to tears

which were no shame to their manhood.   The following
letter refers to this visit.

To Miss Burton

' 3 Queen's Road Villas, Queen's Road, Battersea,
London, Sunday, October 5, 1877.

' Dear Bess,—I guess I am a considerable delinquent in
the matter of correspondence, but I have many excuses,
which I trust you will accept on credit.   I have had very
little spare time since I have been back, work at Thames
being heavy, and the getting out to Battersea long and
tedious matter, consuming nearly three hours a day.   I
had six days in the north, while the Iron and Steel
Institute had meeting at Newcastle.   We went all over
the place with special trains, and saw the Works of the
place to our hearts' content, and wound up by a walk from
Middlesbrough to Whitby.   I have been reading Mac-
aulay's Life—quite charming, but one doesn't know which
most to admire; his stupendous mental capacity, including
the vastest memory mortal ever possessed, or his character
as a man.   I have embodied your finance into a condensed
addendum.   I wish you would check everything directly
you get it, as I keep no memorandum of your transactions
beyond what I send you.—Yours,

' SIDNEY G. THOMAS.'

Meanwhile Gilchrist was now fairly infected with belief
in his cousin's theory, and was working away with a will.
In the rough shed on the Welsh hillside many scores of
' blows ' were made with the greatest energy and enthusiasm—' blows ' chiefly conducted in the late evening or
night, for the Blaenavon analytical chemist had naturally
to work in secrecy in his leisure hours.   On October 19,
1877, Gilchrist writes to Thomas :—

'I want you to come down that we may get some experiments made. I can manage the analyses all right;
but I should like your assistance in the experiments—so
say you will come.'

About this time, as letters of this sort arrived, and
good news of successful results, there began for Sidney a
new phase of anxious and feverish activity. He found it
indispensable to be on the spot at Blaenavon, and this was
only possible by means of hurried trips to South Wales in
days snatched from his regular avocations at the Thames
Police Court—days which had to be reimbursed, so to
speak, by extra toil at other times. He would often go
down by midnight train on a Thursday night, and return
only just in time for court on the following Monday
morning. He had always, as his cousin has told us when
speaking of the French tour in 1869, been habitually
careless of needful nutrition and rest, and in these months
he became more careless than ever. The constant letters
to Mr. Gilchrist, some of which we have quoted, were
generally written from Arbour Square during the midday
adjournment which should have been devoted to a meal;
but Thomas still, despite remonstrance, cherished his view
that lunch was a superfluity. The strain of anxiety and
labour, the midnight journeys and the life at high pressure
called urgently for double fuel to be supplied to the
machine; but the demand was too frequently disregarded.
There is no doubt, unhappily, that at this time, when a
great triumph of vast importance to the whole world was
in preparation, there were developing also the seeds of the
malady which was to cut short in but a few years more a
bright and really glorious career. Grave mischief was
especially wrought by a long run along a railway line to
catch the train back to London. The strain on the lungs

was too much for the over-worked and under-nourished frame, and manifested itself by a sudden fainting-fit and fall. To this strain on the lungs may perhaps be ascribed the 'emphysèma' which was eventually set up, and which little more than seven years afterwards resulted in a death premature indeed.

The contributions to 'Iron' were, meanwhile, still going on, no complication of work seeming too much for Sidney's eager and indefatigable spirit. On November 3, 1877, he writes to Mr. Gilchrist :—

'I went to Chemical [Society] the other night. Awfully slow. To my intense surprise, Vallentine came up to me and paid me an elaborate compliment on my *ferric* essays.'

An additional field of work, which absorbed an immensity of time, was contemporaneously opening out— Patent Law, both British and Foreign, had to be studied, and where Thomas was the student, study meant exhaustive study.   British Patent Law is by no means simple, and in 1877 was probably less simple than now ; but Foreign Patent Law is frequently troublesome indeed to an Englishman.   Sidney mastered the whole subject in all its branches, his legal training, although in so different a field, being doubtless of advantage to him.   The gentleman who afterwards became his patent agent and a valued personal friend as well, testifies that he has learnt much Patent Law from him.

Beyond investigating the law on the subject, the records of the Patent Office had naturally to be searched, that full knowledge might be gained of what had already been done in the direction of dephosphorisation.

Towards the end of November Thomas writes to Wiesbaden :—

To Miss Burton
'November 22, 1877.

'Dear Bess,—All best wishes for so long a succession of 24th's as you may wish to enjoy, some, I hope, with us; but, if not, wherever you may be, may you be happy. I had the idea of writing you a long letter for the 24th; but a week ago some experiments in iron metallurgy in which I had been long occupied came, under Percy's care, to a sufficiently successful issue to have kept me ever since at the Patent Office for every spare moment. I am afraid it won't bring any fruit but anxiety; but the result is satisfactory, nevertheless, as confirming theoretical deductions I had arrived at by much toil.

'I am due now and overdue, so, with all best greetings,
'Yours ever,
'SIDNEY G. T.

'You will accept my intentions as equivalent to the longest and pleasantest letter I have the [power] to scribble.'

On November 23, 1877, Thomas writes to Gilchrist:—

'Your letters are the events of the day. Though I have less to record, I have not been quite idle. I have hunted up every specification that abominable indexes for past ten years give any clue to.'

Later in the month he writes:—

'I have been asked to go down to Cwm Avon as Commissioner for dissatisfied shareholders, to investigate sale. I don't think I shall. As you are known so well there, it might be unpleasant to you.'

This last note illustrates both his careful consideration for others' feelings and the confidence that was already placed in this still unknown young man of twenty-seven by those who had come in contact with him. In this month of November, a busy month indeed, Thomas actually did take out his first patent, although the complete specification was not filed until the following May.[1]

On December 3, 1877, he writes to Mr. Gilchrist:—

'I have told Chaloner not to expect anything from me but one article I had promised, and which will bring in a little coin, of which I am anxious to secure and save all I can for "the cause." I have therefore nothing but translations and revisions, which don't take long, to divert me. Unfortunately Thames is progressing very fast in severity of work. We get now nearly a thousand convictions a month, besides a multitude of cases which, though investigated at length, result in acquittal or dismissal.

' If additional coin will hurry up construction of blast engine do not scruple to use it. You must have worked tremendously to get such a magnificent crop of results. Take care of yourself. Have had two and a half hours' interview with Patent Agents.'

Thomas, however, amid all these occupations found time to send Christmas greetings to his cousin in Germany :—

To Miss Burton

'Dear Bess,—All good wishes for the 25th and still more for the first and all other days of '78. I am I fear a hopelessly bad correspondent just now. The epidemic of invention has found me an easy victim and possessed me body and soul, though not to the eternal exclusion of all

---

[1] Events, however, moved so quickly that in July of next year (1878), and long before his discovery was generally known, Thomas says 'I regard this patent as somewhat out of date,' and, in point of fact, patent succeeded patent down to the day of his untimely death.

thoughts of the absent. I have now nearly finished reading the 500 and odd specifications of my predecessors in the field, "all of whom have failed," and I have made suitable arrangements to add my bones to theirs, though I am just now tied up for want of immediately available funds. My first trial comes off in January down in Wales, some experiments on a small scale having given results remarkable in a scientific point of view. The problem is the separation of phosphorus in the manufacture of Bessemer and Martin steel.—Yours,

'S. G. T.'

However, the specifications of former adventurers in the same field were gone carefully through a second time; for on January 29, 1878, he writes:—

'I have gone through the last twenty-two years' specifications again with Lily's help.'

At the end of 1877 and the beginning of 1878 the results of the experiments which had been continued for now something like nine months with constant energy and zeal had proved thoroughly satisfactory. After trials in crucibles, a miniature converter had been obtained, which, although it only held eight pounds, instead of eight tons, sufficed for experimental purposes. Soon after Sidney's return from abroad, Northampton pig-iron had been partially dephosphorised by lining the converter with bricks of limestone and with silicate of soda. For some time, however, from some defect in the apparatus, the experimentalists were not able to get a cast fluid, so as to finish the operation. Later in the year complete success was achieved, still of course upon the miniature scale; and they obtained a number of casts of eight pounds each, which upon analysis were found to be excellent steel.[2]

[2] *Creators of the Age of steel,* by W. T. Jeans, London, 1884, p. 305.

The old difficulty of inventors was, however, rising as
an obstacle in Thomas's path, the difficulty of finance. In
his case, although the difficulty existed, it was minimised,
partly by his own wonderful frugality and forethought,
partly because he was fortunate enough to meet, not with
the typical capitalist, but with just and straightforward
men. Thomas had contrived during his ten years' servi-
tude at the Police Court to save out of his not too
abundant salary [3] no less than 800*l*., which was to be
devoted to 'the cause.' It was a large sum for him at
that time; but expenses were heavy and he was becoming
anxious as to what would happen when it should be
exhausted. He was determined not to accept the offers of
further supplies which were made to him by his mother
and by one or two family friends who knew he had a big
scheme on hand.

For this reason, therefore, if for no other, an event
which happened in the earliest days of 1878 came in good
time.

The manager of Blaenavon Works, Mr. Edward Martin,
said to Mr. Gilchrist, 'I know you young men have some
secret work on hand. I think it would be well if you put
confidence in me.' Confidence *was* put in him and Mr.
Gilchrist's analyses were submitted to him. Mr. Martin
was so much struck with the basic theory and the proofs
afforded of its truth that he at once afforded facilities for
further experiments at Blaenavon on a larger scale and
obtained for the ' young men' promises of similar facilities
at the Dowlais Works, of course upon terms favourable to
the two companies should the process continue to succeed.
He also undertook personally to purchase a share in the
patent.

Thus the financial difficulty was removed. Moreover,

[3] See *ante*, p. 12.

the adhesion of a clever, practical, business man to the process was in itself an immense moral support.

From that time forth Thomas had to the last day of his life Mr. Martin's loyal co-operation, the loyal co-operation of a whole-hearted friend and ally whose word was his bond. Such help could not fail to be in itself a great pleasure to him who was aided by it. Mr. Martin, having committed himself to the enterprise, threw himself into it with characteristic energy, and his suggestions and experience were found to be invaluable.

The adhesion of Mr. Martin gave an immediate impetus to the investigation, and the promised experiments were at once carried out both at Dowlais and Blaenavon. At Dowlais the trials were not entirely successful for reasons which will appear presently; at Blaenavon they were continued with satisfactory results throughout the spring and summer. Thomas shall describe them presently in his own words.

Shortly before the Dowlais trial, Thomas writes to Wiesbaden :—

To Miss Burton

'3 Queen's Road Villas, February 20, 1878.

'Dear Cousin,—Your letter was a very pleasant one to me. I should have written you some weeks since had I not been pressed on all sides for time. Last week was down at Blaenavon for three days to coach my pet through some infantile disorders. We are a long way yet from a commercial success, though the indications are very favourable. I arranged while in Wales for the Dowlais Works, the largest in the world after Krupp's, to give me a big trial in a month. After that I shall be more clear as to my chances. Percy has been working hard as to details and analysis.

'I am thinking of plunging into foreign patents to the

amount of 100*l.* or so next.   Money is a commodity which goes but a small way in these matters.

'I too have been eager in politics of late.   I should be exasperated if we blundered into a senseless war.   The danger is now much more remote than it was last week, when we hourly expected a collision.   Going down to Wales I travelled with an intelligent man who had been much in India, Bosnia, and the Danubian Principalities.   We had much talk, from which I gained more information than from a legion of articles.   He by the way writes for the "Nineteenth Century," which, with the "Contemporary" and "Fortnightly," represent the cream of modern thought.

'I've had a note from Percy this morning of more difficulties encountered ; I shall have to go down to see them, I expect.   My light reading now is Patent Law, most contradictory of studies.'

In March, however, the first public announcement of the new process was made, although the announcement attracted no particular attention.

At the spring meeting of the Iron and Steel Institute, Mr. I. Lowthian Bell read a paper on the separation of phosphorus from pig-iron in a furnace lined with oxide of iron.   The whole question of dephosphorisation was discussed by several speakers, amongst others by Mr. Snelus. At the end of the discussion Thomas, who was present as a visitor and who was probably the youngest man in the room—who certainly with his clean-shaven face looked the youngest—managed to get an opportunity of utterance. His words have been preserved and show a characteristic quietude of phrase.   He said :—

'It may be of interest to members to know that I have been enabled, by the assistance of Mr. Martin at Blaenavon, to remove phosphorus entirely by the Bessemer converter. Of course this statement will be met with a smile of incredulity, and gentlemen will scarcely believe it; but I have the results in my pocket of some hundred and odd analyses by Mr. Gilchrist, who has had almost the entire conduct of the experiments, varying from the very small quantity of 6 lbs. up to 10 cwt., and the results all carry out the theory with which I originally started and show that in the worst cases 20 per cent. of phosphorus was removed, and in the best I must say that 99·9 was removed; and we hope that we have overcome the practical difficulties that have hitherto stood in the way.'

Mr. Chaloner, who was at the meeting, described long afterwards in 'Iron' (February 6, 1885) the reception given to this declaration. 'We well remember the sneer as well as "smile of incredulity," which spread over that meeting, and can testify to the scarcely veiled antagonism exhibited to the unknown youth who had presumed to proclaim the solution to a problem which the leaders of metallurgy had pronounced well nigh insoluble.' No observation of any kind was made by anyone.

We need not be angry with the assembled experts. Their attitude is probably very fairly described and explained by Mr. Jeans. 'The meeting did not laugh at the youthful *Eureka*, nor did it congratulate the young man on his achievement, much less did it inquire about his method of elimination. It simply took no notice of his undemonstrative announcement.' [4]

Thomas went on quietly working with the aid of Mr. Martin and his cousin at his experiments. He was, as appears by the following letter to Miss Burton, by no means displeased at provisional absence of interest by scientists in general. This letter, too, brings out strongly

[4] *Creators of the Age of Steel*, p. 303.

the estimation in which Thomas was held at the Thames
Police Court by the magistrates under whom he served.
No external occupations, however engrossing, ever interfered, we cannot too often repeat, with his zealous and
whole-hearted discharge of his official duties :—

'Thames Police Court : April 8, 1878.

'My dear Bess,—I have had to send your Italians to
Florence for fresh coupon sheets, as old ones exhausted.
. . . My experiments are rather at a standstill. Some
great Works promised me a trial two months ago; but
have not made the necessary preparations yet.

'However, nearly 300*l.* has been spent in patents, in
anticipation of things turning out well.

'I said a few words on the discussion on Bell's paper;
but we wish to keep quiet at present. I forget whether I
told you of the sudden death of my colleague as he was
returning to the office after a short holiday . . . His suc-
cessor has only just come, so I have been over full of work.
The Magistrates went down to the Home Office on their
own account, to try and get the rule of seniority set aside
in my favour, which was rather gratifying. Of course they
were unsuccessful. . . .

'Here the east wind is on the rampage, and has knocked
up most people.

'I utterly abjure all breath of war and slaughter, and
am utterly ashamed of the miserable position we have
blundered into. The Russian may be as black as he is
painted, but neither he nor we will be improved by
slaughter.—Yours always,

'S. G. THOMAS.'

The next two letters to Germany give further glimpses
of the many cares pressing on the restless and indefatigable
mind of the writer.

I

To Miss Burton

'May 19, 1878.

'Dear Bess,—A friend of Lil's, whom I think you know, wants to get languages with a view of getting a better engagement.

'They are three orphans, and coinless nearly. She has been over here to-day, proposing to go to Paris on Miss H.'s recommendation. I suggested she would do better in Germany, to which she assents. Now could the B.'s take her? It seems she would about fill the vacancy for which your advertisement was. She is I am told about twenty-two, has been three or four years teaching, and would be willing to pay something. If the vacancy is filled up, as from your last you seem to think probable, what would you advise? Do you know of anything else? She knows no German, can teach English well, can't pay more than 25*l.* per annum. The mother is very anxious to do something for her. I should think lots of German families would like to get an Englishwoman to teach for nothing. I am up to my ears still in patents. I shall have a hard fight, but even if beaten, fighting does one good. I have not heard yet if they have granted my German patent. They refuse a great number. I go down to Wales again in a week, and hope to do something on the big scale. Have had to go to the Opera twice lately; *Ruy Blas* last night, *Tannhäuser* a fortnight ago. I was dreadfully bored by both. I have an impression that I used to enjoy the two or three times I went with you. We have been reading Heine's life, very interesting, discursive on German and European literature and politics. Have now the third volume of "Prince Consort's Life," which of course has especial bearing on the policy of the day. I do not think you would gain anything by selling South Italians unless

at a high price. It is almost impossible now to get a decently safe 6 per cent. Still more difficult in Germany.
Please answer by return as to your opinion on the second
question.—Yours,

'S. G. T.'

'July 20, 1878.

'Dear Bess,—I don't know if you or I am the worst
correspondent, but I think if you knew how I was driven
you would absolve me with honour for all my failings
therein. Phosphorus is a subject which engrosses an incredible amount of time. My visit to South Wales showed
that while scientifically my views are entirely confirmed,
there is much money (some thousands) to be spent in
putting things on a fair technical footing, and much more
in legal defence of my position. As I do not possess
these thousands, I am not going to bother myself about
trying to force my views commercially, but let them rest
with doing what I can to establish them, for the benefit of
people at large. I am now fighting Krupp of Essen and
the Bochum Steel Co. As they write their objections in
German, and require to be confuted from German authors,
this is not easy. So I shan't see you in Paris, whither I
hope to go for a week or two in September. I hope you
will have a pleasant holiday in the Schwarzwald. I saw
your last protégé off on Saturday. It made me think I
should like to run over.'

A day or two after Thomas writes to his sister, who
was away from home on a visit :—

'Thames Police Court, London, E.: July 22, 1878.

'Dearest Child,—The mother flourishing and dashing
about all over country. Being free from surveillance, I
am increasing in weight daily, through the adiposing effect

I 2

of a peace and quietness which I don't always enjoy. You seem to be leading a "jollies" (not jolly) existence, which ought to do you a world of good. Don't go drowning yourself—not too frequently. I enclose as a matter of benevolence something for you to do to fill up the vacuity of your existence. Will you on enclosed ruled paper make *two* copies of also enclosed results as neatly and legibly as you can, and let me have them back not later than Thursday morning, and receive my blessing? I have put in two or three to show how I should like them done, only neater. Use your sense in locating remarks, &c., and leave spaces when clearness improved thereby. I am over ears in work. Krupp of Essen, and another, are attacking me in German, and I have to refute them by German authors. Fighting with your head in a bag is a trifle to it.—Yours,

'S. G. T.'

CHAPTER X

THE BASIC PROCESS DESCRIBED

DURING this summer Thomas in collaboration with Mr. Gilchrist wrote for the approaching autumn meeting of the Iron and Steel Institute a paper on 'the Elimination of Phosphorus in the Bessemer Converter.' We cannot do better than give here the substance of this paper (omitting technicalities and distasteful figures as much as possible), since it furnishes the results of the experiments and describes the point at which the process had arrived and its *rationale* in the words of Thomas himself.

'The non-removal of phosphorus in the Bessemer Converter,' write the authors, 'owing to which the great bulk, not only of British, but of French, German, and Belgian ores are still unavailable for steel-making, is a fact too familiar to metallurgists to need insisting on. The inquiry whether this unfortunate circumstance is due to causes absolutely inseparable from the conduct of the Bessemer process, or to others which are merely the accidents of a particular mode of constructing the apparatus, is obviously of vital importance. If the non-elimination be due to the intensity of the temperature or to the short duration of the operation, or to both these causes combined, it is almost hopeless to expect that we shall ever be able to use ordinary unpurified pig-iron in the Converter.

'That it is to these essential accompaniments of the process that the phenomenon of the retention of phosphorus by Bessemer metal is to be ascribed, is—it is believed—the generally received opinion and one which has comparatively recently received the sanction of the weighty authority of such eminent metallurgists as Mr. Lowthian Bell, Dr. Wedding, Professor Kerl, and M. Euverte.

'An examination of the general conditions attending the removal of phosphorus in puddling and refining operations taken in connection with the well-known action of silica on phosphate of iron at high temperatures, and the fact that in many other processes in which the temperature is very high the elimination of phosphorus is not apparently effected, seems, however, to justify the belief, which may have probably suggested itself to other members of the Institute, that it is to the silicious lining of the ordinary converter and to the consequent necessarily silicious quality of the slag, that the one defect of the Bessemer process is due.   Under this conviction, at all events, experiments were commenced by the authors about three years ago on the effects of basic lining and basic additions in the several steel-making processes.   Unfortunately the appliances at command were of a very imperfect character, and the results obtained, though highly encouraging, were —owing to defects in the miniature Converter employed, which prevented our ever completely finishing a blow—not entirely conclusive as to commercially complete purification being possible.

'While awaiting the completion of an improved Converter which was unavoidably delayed for some time, we were encouraged by finding that M. Gruner, the distinguished professor of the Ecole des Mines of Paris, laid great stress on the silicious character of the cinder and lining in the Converter.   M. Gruner, however, seems at that time

to have regarded this as one only of three causes which
prevent elimination of phosphorus, and proposes as a
remedy the preliminary refining of phosphoretic pig before
it is attempted to convert it.

'With a new Converter, a large number of experiments
were made in the autumn of last year, which gave much
more definite results. The lining used in these experiments consisted of limestone and silicate of soda, a mixture
which had been found to answer well in earlier trials. . . .

'On laying some of the first results obtained from
this 6 lb. Converter before Mr. Martin of Blaenavon,
he at once recognised their importance, and from that
time we have been deeply indebted to him for his un-
failing and liberal support and much valuable advice and
assistance.

'The Blaenavon Company without hesitation undertook
to put up apparatus to carry the experiments further, and
has with great spirit fulfilled its promise to test the value
of the theories thoroughly.

'In a vertical Converter, taking from 3 to 4 cwt. of
metal, results confirmatory of those previously observed
were obtained.   In the six-pound Converter liquid decar-
bonised iron could not be obtained ; but in the new vertical
Converter this was readily done. . . .

'Some fifty or more blows were made in this vertical
Converter, and the products analysed ; and it was found
that, using a basic lining, it was generally necessary to
continue the blow for about forty seconds after the flame
dropped in order to bring down the phosphorus very low.
With this proviso, the elimination of phosphorus could be
secured with absolute certainty.   With a silicious lining
the retention of all the phosphorus in the metal was, as
usual, equally invariable—even when, as in Mr. Bell's experiments, the blow was continued till a considerable pro-

portion of the iron was oxidised. At the same time more
phosphorus and less silica would be found in the slag
obtained under these conditions than appears to be the
case when large quantities of metal are treated under
similar circumstances. . . .

'It would seem that the presence of a considerable
amount of lime in a not too silicious slag is highly favourable and on a large scale essential to the removal of
phosphorus. As it was manifest that phosphorus was not
removed until the slag was sufficiently basic, the effect of
large basic additions in combination with a basic lining
was tried. With the object not only of obtaining a highly
basic slag at an early stage of the blow, but of rendering
the operation independent of the wear of the lining by
which alone the basic character of the slag is otherwise
obtained and maintained, advantage was taken of the fact
that lime and oxide of iron are fusible in many propor-
tions. . . .

'With a 12 cwt. Converter of the ordinary pattern, ex-
pressly put up by the Blaenavon Company, only a limited
number of casts have been made, owing to a deficiency of
blast. . . .

'By the kindness of Mr. Menelaus, for whose invaluable
assistance we tender our warmest thanks, we were enabled
to try, at the No. 3 Pit at Dowlais, if the superior intensity
of heat which might be expected from the conversion of
five or six tons of metal at a time affected the conclusions
to which smaller experiments pointed. It was intended
to line this Converter with highly burnt basic bricks.
The bricks intended for this purpose were, however,
accidentally under-burnt, and so spoilt, hence recourse was
had to a rammed lining of limestone and silicate of
soda. . . .

'These results appear to confirm the conclusion that

for the process to be of technical value, waste of lining must be avoided by making large basic additions, so as to secure a highly basic slag *at an early stage of the blow.* In these trials, however, it was thought prudent to feel our way, and not add at once the very large amount of base which our theory demanded, the more so as we were not able to add the bases in a heated state. It is also made clear that a slag containing under 14 per cent. of iron may be very effective in removing phosphorus. . . .

'It is obvious that without a sufficiently durable as well as refractory basic lining, the simultaneous dephosphorisation and conversion of cheap pig in the Bessemer vessel cannot rank as a commercial process. Our early experiments rendered it clear that ordinary nonsilicious lime and limestone did not constitute by themselves a satisfactory lining material, nor were renewed trials, made after becoming acquainted with a patent dealing with their application, more successful; magnesia, the use of which as a furnace lining has been suggested by M. Caron and others, is at once very expensive and, when used by itself, very tender. After a very extended series of trials it was, however, found that by firing bricks made of an aluminosilicious limestone at a very intense white heat, a hard and compact basic brick is formed. These bricks unfortunately labour under the defect of a liability to disintegration when exposed to the action of steam. By the use of certain aluminous magnesian limestones and equivalent combinations, and an otherwise similar mode of manufacture, this difficulty has been, after many failures, overcome. . . .'

Here we have the problem clearly stated, namely : 'The simultaneous dephosphorisation and conversion of cheap pig in the Bessemer "vessel," in such fashion as to make the process a commercial success.' The problem is solved by substituting a reasonably durable basic lining for the former silicious, and therefore acid one, and by avoiding 'waste of lining, by making large basic additions, so as to secure a highly basic slag *at an early stage of the blow.*'

CHAPTER XI

TRIUMPH

ANXIOUS as these times of waiting were, while this paper
was being written and the experiments continually
watched (the regular toil at Thames Police Court still
going on), it is characteristic of Sidney that he should have
found time to take lessons in French conversation. Regularly for three months he was an hour late for dinner every
other day, nor was any explanation obtainable by his
relatives for a long period. The real explanation was that
he would stop in the City on his way from Arbour Square
to Battersea (where, it will be remembered, the family
were now dwelling), to have an hour's educational talk
with an old Frenchman. It was only later, when all were
gathered in Paris, that upon being complimented upon
his fluent Gallic speech, he revealed the little secret.

In September the autumn meeting of the Iron and
Steel Institute was held in Paris—held there, especially,
because of the Great Exhibition of 1878. Thomas arranged
his annual holiday from his official duties to coincide with
this meeting, and went to the gay city in company with
his mother, sister, and a friend. Mr. Gilchrist also
attended. The paper on the 'Elimination of Phosphorus'
was put down for reading, and originally placed near the
top of the list; but belief in the alleged discovery of an
unknown youth had not much spread since March, and
the paper was removed to the end, and then left by the
authorities unread for 'lack of time'; a course not

altogether disagreeable to Thomas, who was anxious to
further secure the patent position. This action attracted,
however, some attention—especially as a portion of the
paper had appeared in 'Engineering' before news of the
change of programme could reach that journal. Moreover
the paper was freely distributed among members. But
even if the non-reading of it had been a great disappointment, there would have been ample and unlooked-for
compensation.

Thomas accompanied other members upon an excursion to the great Works of Creusot, and there, as good luck
would have it, fell upon talk with Mr. E. W. Richards, the
manager of Bolckow, Vaughan, and Co.'s huge Works in
Cleveland. Sidney's remarkable personality, and vivid,
lucid discourse never failed to impress those with whom he
came in contact; and Mr. Richards proved no exception
to the rule. Cleveland, it must be remembered, is the
district in all England which suffered most from the non-
elimination of phosphorus in the Converter; for the whole
of its ores (and it had an annual output of 6,500,000 tons)
were phosphoric, and, therefore, as was then thought,
useless for making steel by the Bessemer process. Naturally, the conversation turned upon the alleged discovery
which was to change all this. Thomas explained to Mr.
Richards the position in which the experiments stood, and
the desire that was felt to continue them on a larger scale.
A meeting was arranged to discuss the matter further, and
it is not too much to say that the further discussion at that
meeting secured the immediate commercial success of the
process.

Mr. Richards had better tell the story in his own
words : [1]—

[1] Words taken from Mr. Rich-   Cleveland Institution of Engineers
ards's presidential address to the   (November 15, 1880).

'Messrs. Thomas and Gilchrist prepared a paper, giving very fully the results of their experiments, with analyses. It was intended to be read at the autumn meeting of the Iron and Steel Institute at Paris in 1878; but so little importance was attached to it, and so little was it believed in, that the paper was scarcely noticed, and it was left unread. . . . Mr. Sidney Thomas first drew my particular attention to the subject at Creusot, and we had a meeting a few days later in Paris to discuss it, when I resolved to take the matter up, provided I received the consent of my directors. That consent was given, and on October 2, 1878, accompanied by Mr. Stead of Middlesbrough, I went with Mr. Thomas to Blaenavon. Arrived there, Mr. Gilchrist and Mr. Martin showed us three casts in a miniature cupola, and I saw sufficient to convince me that iron could be dephosphorised at high temperature. I also visited the Dowlais Works, where Mr. Menelaus informed me that the experiments in the large Converters had failed owing to the lining being washed out. We very quickly erected a pair of 30 cwt. Converters at Middlesbrough, but were unable for a long time to try the process, owing to difficulties experienced in making basic bricks for lining the Converters and making the basic bottom. The difficulties arose principally from the enormous shrinkage of the magnesian limestone when being burnt in a kiln with an updraught, and of the failure of the ordinary bricks of the kiln to withstand the very high temperature necessary for efficient burning. The difficulties were, however, one by one surmounted, and at last we lined up the Converters with basic bricks; then, after much labour, many failures, disappointments and encouragements, we were able to show some of the leading gentlemen of Middlesbrough the successful operations on Friday, April 4, 1879. The news of this success spread

rapidly far and wide, and Middlesbrough was soon besieged
by the combined forces of Belgium, France, Prussia,
Austria, and America.   We then lined up one of the six-
ton converters at Eaton and had fair success.'

Meanwhile Thomas, while following and taking part in
these anxious experiments, thus finally crowned with definite
triumph, had not been idle in other directions.   His
energies had been devoted to safeguarding the patent
position at home and abroad as though he had no other
work on his hands.   The patent of 1877 had been rapidly
followed by other dephosphorisation patents of January
1878, March 1878, and two of October 1878. Other patents
were taken out in 1879.   In January of the latter year
two patents were taken out for basic bricks, and a series
of patents for treatment of slag begin in November 1878.
In foreign countries the same activity was displayed.

Thomas in the following letters gives us some glimpses
of his proceedings during the period between September
1878 and April 1879—the period which assured the commercial success of his process and which has just been
described by Mr. Richards.

To Miss Burton

'Thames Police Court :   October 3, 1878.

'Dear Bess,—I was so sorry you did not make your
appearance in Paris.   I had quite looked forward to it and
had the impression you had promised it.   The fortnight
spent there was most enjoyable, the weather beautiful, the
city ditto, and the Exhibition magnificent.   I went to the
Exhibition seven times and only saw half imperfectly.
The Mother was happy all day long and our quarters excellent and, considering the prices current, not dear.   I
think the city much improved since I saw it in '69 with

you and Robert. I spent my first three nights on the sixth floor of a queer old inn close to our old quarters; this time in the Rue Montmartre. Our paper was postponed,—the preference being very properly given to foreign papers, and the course adopted suiting us very well. It still occupies a great fragment of my attention. I returned to England last Friday and have been living à la Crusoe in the empty house. . Tuesday night, however, I had a telegram which sent me down to Wales by the mail, to meet some great North of England guns who had come to Blaenavon to see our experiments. They were well impressed with what they saw and I returned last night. . . . I shall probably be in Belgium to try to start some Works there early in next month. On the whole my hands are pretty full. Whether we shall succeed in getting any pecuniary advantage remains to be seen, I am afraid of the funds which are a necessity for victory being wanting. However, of exciting employment it seems we shall have enough.—Yours always,

'S. G. T.'

'I will send you a copy of our paper when I can.'

'November, 1878.

'I am awfully busy, or should have written you before this. Things are *in statu quo*, but I am much more occupied. I go to Belgium to-morrow to superintend some experiments. I shall have rather a cold time of it.'

'December 23, 1878.

'Dr Bess,—Yr letter found me at Middlesbro', where I think things are progressing fairly. Percy was with me, he is director of all practical details, and works like—anything. His Co., the Blaenavon Iron Co., have just failed. He doesn't know yet how it will affect him, but it can

hardly fail to be detrimental. It is also very unfortunate
for our patent interests, as the Co. had engaged to give
us a large trial at once. From Middlesbro' I went to
Cumberland, and came home by mail last night, being
from 7 P.M. to 9 A.M. on the road. I was nearly frozen.
I am getting rapidly ruined, but having plenty to do
induces me to regard the contingency with equanimity.
We shan't know how we stand for another six months at
least . . . There is a terrible amount of distress throughout England . . . My Belgian visit was quite enjoyable
and the result on the whole quite satisfactory, *i.e.* fairly
good steel to the amount of seven or eight tons made from
stuff that had never made steel or anything like it before.

'I see a good deal of Americans just now. I have
struck up an alliance with one I encountered abroad, and
had to stay a few days to the home folks' amusement.

'Ever y^{rs} always,

'S. G. T.'

When the news of the experiments of April 4, 1879,
spread abroad, would-be users of the process on the
Continent found themselves face to face with the patent
rights which the forethought of Thomas had secured. A
literal race to the quiet home in Queen's Road, Battersea,
at once began. The present writer well remembers Thomas
telling him, with some glee, a curious story of the eagerness of foreign ironmasters to secure licences, a story
which is also a sermon on the text of striking while the
iron is hot. One April Sunday night, two Belgian steel
manufacturers from the same neighbourhood crossed
together in the same boat. M. A—— and M. B——
conversed the whole way, but neither said a word of their
errand to Albion. They both drove to the Royal Hotel
on the Embankment, upon their arrival at Charing Cross

at some unearthly hour on Monday morning. M. A——
thought he might safely go to bed for a couple of hours
and then have some breakfast before pursuing his journey
to the wilds of remote Battersea. M. B—— was wiser
in his generation; he chartered a hansom directly he had
shaken off his fellow-traveller and rang up the quiet household in the Queen's Road at 7.30 A.M. He secured an
audience with Thomas and proceeded to negotiate terms
for the use of the process. The interview lasted for three
hours and was just concluding, when a telegram arrived
from M. A—— announcing that he was on his way. At
noon he duly arrived, congratulating himself on his
promptitude. Alas! M. B—— had secured the monopoly
of the process for the district.

It is probably to this Belgian arrangement that allusion
is made in the following letter :—

To Miss Burton

'3 Queen's Road Villas, April 12, 1879.

'Dear Bess,—Many thanks for your congratulations
of 10th. Of your sympathy I of course felt myself sure.
It is, however, not the less pleasant to receive them. Yes,
after some work, we have solved the greatest industrial
problem of England; so at least people who have been
themselves trying the solution for twenty years say.

'We have certainly secured some *reputation*, and may
(or may not) secure some money.

'This last we shall know in two or three months, but
not before. Till this is ascertained I do not want to give
up Thames, as I have to spend about 50*l.* a month still
on one thing and another. Of course I pay all Percy's
extra expenditure. I have just concluded an arrangement
with some Belgians, and shall probably have to take a
continental trip in a few weeks. You may imagine I am

K

pretty busy; I spent three nights out of six on the rail last week.—Yours,

'S. G. T.'

The deferred paper of Thomas and Gilchrist was duly read at the spring meeting of the Iron and Steel Institute which was held in London. 'That meeting was,' says Mr. Richards, 'perhaps the most interesting and brilliant ever held by the Institute.' Mr. Bessemer (not yet Sir Henry) came forward with a cordial recognition of the new and wide-reaching development of his epoch-making process. 'Phosphorus,' he said, 'has been my difficulty and my bane.' If it had not been for discovering that steel could be made from Swedish pig without the necessity for dephosphorisation, he might have continued on the road he had entered upon. 'Whether I should have arrived at the results which the present inventors have arrived at I cannot tell. . . . I hope and believe they will be able to receive the recompense which their talents and industry deserve.' [2]

'Directly this meeting was over,' says Mr. Richards in the presidential address already quoted, 'Middlesbrough was again besieged by a large array of continental metallurgists, and a few hundredweights of samples of basic bricks, molten metal used and steel produced were taken away for searching analysis at home. Our continental friends were of an inquisitive turn of mind and, like many other practical men who saw the process in operation, only believed in what they saw with their own eyes and felt with their own hands. And they were not quite sure even then, and some are not quite sure even now (1880). We gave them samples of the metal out of the very nose of the Converter.'

On May 10, 1879, Thomas resigned his junior clerk-

[2] *Iron*, May 17, 1879.

ship at the Thames Police Court, after nearly twelve years
of service—service as energetic as if his duties there had
been the sole object of his life. We have seen (*ante*, p.
24) what Mr. Lushington has said upon this point. Thomas
left nothing but good wishes behind him. The constant
drain upon his energies,—otherwise fully, more than fully
occupied,—must (especially during the last three anxious
years) have been serious indeed. Yet daring as he was
(often indeed seemingly reckless), it was very characteristic
of him that he did not abandon this modest certainty
until the path to fortune was clear before him. Neither
the acceptance of the new process by Mr. Martin nor its
adoption by Mr. Richards was sufficient to induce him to
burn his boats behind him ; it was not until continental
ironmasters were competing for concessions that he made
up his mind definitely to break with the Civil Service.

Let us say, once for all, here, that the Sidney Thomas,
the triumphant inventor, was in every respect the same
Sidney Thomas he had been years before, when simply second
clerk at Arbour Square—eager, strenuous, and energetic,
but ever preserving the equal mind, and no more puffed up
by victory than he would have been cast down by failure—
always anxious to ascribe success to others more than to
himself.

In the following letter he seems even now somewhat
doubtful of the future :—

To Miss Burton

'May 11, 1879.

' Dear Bess,—We have scored I think *one*. Delivered
paper on Thursday before the largest meeting ever held ;
it was well received by all, both continental and English
metallurgists, and we became *pro tem.* junior lions.

' I have based my foreign patents nearly all on terms

which *may* pay us well, and I hope we shall eventually do
some good business in England, though they are much
behind their continental rivals in enterprise.

'We were introduced to everyone, and the effect of
the whole is by no means disappointing. Even Krupp's
engineer paid us high compliments. I have done my
best to give the Phœnix Works a good chance, though
German patents are largely out of my hands. I resigned
"Thames" yesterday, as I found I could not drive so dis-
cordant a team any longer; so I am now on my own
resources. We have still a lot of new work to go through,
and not a few risks to run on account of the magnitude
of the stake. Whatever happens, I think we have been
fully rewarded for our work. Of course I *have* your congratulations; you had better come and bring them.'

The resignation at Thames brought little relief to his
incessant labour; the vacant hours were instantly filled by
other toils. The whole of the negotiations for his foreign
patents fell to him to conduct. In some countries and
districts he sold his rights; in others he conceded licences
to individual ironmasters; in others, again, he appointed
agents to receive royalties. The basic process spread with
the greatest rapidity on the Continent, where phosphorus
had been even a more formidable foe to steel-making than it
had been here. Thomas's note-books and account books
during this year show him to have been continually crossing
the Channel, and his striking figure became as familiar in
Westphalian Works as it had been in Arbour Square.

In Germany, however, there was a short but severe
contest with a powerful combination of North German
steel manufacturers. These gentlemen attempted to work
the process regardless of patent rights, and fought the inventor in the law courts, partly on technical legal grounds,

partly on other pretexts. Sidney's letter-book gives a
voluminous correspondence on this matter, and he was also
constantly present on the field in person.

The courts decided in his favour in November 1879.
This in the end, although not, as we shall see, immediately,
settled the question. 'The courts held the validity of the
patents to be thoroughly established, and considered the
substantial novelty and great value of the invention to be
proved and to be such as to amply cover any minor technical defects. This decision was generally welcomed, as
showing that the German Patent Court was determined to
administer the new law on just and equitable principles,
and not on the narrow basis.of the old law, which refused
protection to the inventions of Bessemer and Siemens.' [3]

The following letters refer to this contest :-

To his Sister

'Berlin, November 20, 1879.

'Dearest,—After short conference in my [case], had
two days' dissipation preluding Berlin doing under P——'s
guidance. He is an excellent cicerone. City very fair—
particularly museum; shops brilliant. Went to theatre
in evening—nothing very characteristic—comic opera.
To-day conference; to-morrow and Saturday the fight.
Thirty-six against us. I think we are fairly certain to
lose; but my spirits are good. I shall not forgive you for
neglecting your duty in not having taught me German.
It is a horrible nuisance.

'Look after the mater ! . . .

'Yours very affectionately,

'S. G. T.'

[3] *Creators of the Age of Steel*, p. 314.

To Miss Burton

'Berlin, November 22, 1879.

'Beaten the enemy on own ground.   Sorry I can't call at Wiesbaden.

'S. G. T.'

To his Mother

'Hoerde, November 25, 1879.

'I am visiting at Hoerde.   Spent yesterday morning with Dr. Wedding; also dinner with him on Sunday.   A very jolly little party.   We had great fun.   He was one of my judges; another guest was one of my chief opponents.   Two very pleasant German girls, an American student and an engineer.   They are all coming to stop with us in London for an indefinite period.   By March 1 shall know if I am the proud possessor of 20,000l. or not. The historic name of the family has certainly won notoriety if not distinction.   I am stopping with Massenez.   I leave to-morrow morning for K.

'Spent last morning in Berlin School of Mines, a wondrously perfect place.   Was coached over by Dr. Wedding and an American youth, who regards my humble self as a mirror for aspiring engineers to imitate; but is (nevertheless, or in consequence) a very bright lot.

'It is an awful nuisance not speaking German.   I sat at writing for two days, feeling I *must* get up and make a rattling speech in some tongue known or unknown.

'You will hardly, I fear, hear from me again.   I shall be on the move all along, till Saturday, when I expect to be home for some hours at least.   I have been fed and alcoholised to an appalling extent.   Hope you are taking care of yourself.—Yours,

'S G T'

Meanwhile a difficulty had arisen in this country, which fortunately was at once amicably settled without recourse to litigation. We have said (*ante*, p. 111) that at the meeting of the Iron and Steel Institute in March 1878, when Thomas made his little regarded declaration, Mr. Snelus had also spoken on the dephosphorisation question. This gentleman had indeed had a patent in existence for several years which (it was contended) established the principle of basic linings, although there might be practical difficulties in its application. This patent had been kept alive, but it was not suggested that a ton of steel had ever been manufactured under it. It might also perhaps be said that the many steps in the complete Thomas-Gilchrist process not at all hinted at in Mr. Snelus's specification established a very vital distinction in favour of that process, and indeed that Mr. Snelus's specification had not expressed dephosphorisation as the aim of the patent at all; but it would be both idle and ungracious to pursue a vain discussion of rival claims which both sides from the first treated in a friendly and loyal spirit.[4]

The claims then of Mr. Snelus and of one who became Thomas's valued colleague, Mr. Riley, who had zealously, independently, and ably devoted himself to the lining question, had of course to be considered.

It was agreed to refer to Sir William Thomson's arbi-

---

[4] Mr. Snelus in 1883, after detailing his experiments, said: 'Mr. Sidney Thomas, shortly afterwards, with very much more energy than I had shown, followed in the same line, and Mr. Gilchrist and he developed the process of making basic bricks on a large scale. After this he demonstrated much more publicly than I had done the theory of the basic process, and he induced Mr. Windsor Richards to take it up. It was a piece of very good fortune, I consider, that Mr. Thomas succeeded in enlisting the sympathy of Mr. Richards; this was due to Mr. Thomas's perseverance and to his determination to make the process public and to make it go.'

tration the question of how the profits of the British and
American patents should be divided between the parties—
Thomas being left in sole possession of all continental
rights. Sir William Thomson made his award, an award
ever since cordially accepted and acted on by all concerned, towards the latter part of this year of 1879.

Patents were taken out in America early in 1879, and
led afterwards to much litigation. The quantity of non-
phosphoric iron in the United States is so large, that
probably no country in the world had less need of the
basic process. Yet, as we shall presently see, in no
country in the world was there more interest in the invention and nowhere did Thomas himself receive a more
enthusiastic welcome.

CHAPTER XII

DÜSSELDORF—A GATHERING CLOUD

THE next year of 1880 opened brilliantly indeed for Thomas and the little family of which he was the life and soul. The household gods were in the course of this year removed from Battersea to Tedworth Square, Chelsea, which was Sidney's London home for the remainder of his brief and narrowing span,—a span the narrowing brevity of which was still happily veiled from him and those to whom he was dear. Tedworth Square, however, saw but too little of him ; for most of his time was in this year, as in the preceding one, spent in railway trains, steamers, and English and foreign ironworks.

We have before us many of his post-cards and letters which show something of the intense stress and hurry of his life at this period, and we select a few of them as specimens.

To Miss Burton
'Paris, November 28, 1879.

' Acceptez mes salutations (un peu en retard, je crains, mais pas moins sincères) pour votre *birthday*. All going well, I believe. Shall know how I stand by March 1. Am rather tired, having been *en wagon* two nights. Have two more before me. Heute abends muss ich zurück bis London und dann nach Sheffield, Middlesboro und so weite, so bin ich immer *en route*. The rout of the Teuton, even if only temporary, was angenehm.'

'Liège, February 21, 1880.

'Cara B,—I am toujours en route vous voir, mais toujours si confoundedly pressé que je n'arrive jamais. Wir müssen ein rendezvous en Coeln oder Coblenz haben one day, for a long chat. Have now been Paris, Luxembourg, Hoerde, Ruhrort, Liège, travelling all night (almost every night) and working all day. I had to run through Coeln beide ·Zeite or should have run up to Wiesbaden. It seems dass ich soll nimmer mehr ein jour entièrement libre haben. Your German friends are appealing and causing me a lot of extra Arbeit. Excuse my writing in my ordinary colloquial language, which astonishes some de mes clients. . . .'

'Newcastle, March 13, 1880.

'Dear Cousin Bess,—Though I am, *I expect*, the busier of the two, I am still the best correspondent. . . . As usual, I am wandering over the earth's face. Last week, Sheffield, Blaenavon, Rhymney, London. This week Glasgow, Edinburgh, Newcastle, &c. Hard at it all the time. It is uphill work and complicated; but it is, I trust, to be a big work, and I am satisfied. It is only sometimes rather more than one set of brains can do, to drive so many different horses.

'I think we shall succeed in selling in America for a pretty fair sum. If so, I shall try and secure fair help. . . . . I may be in Germany again in a week or two; if so, I shall try hard to run up to you for a few hours, but I never get nearer than Coeln, and am always driven even for an hour. . . .

'I have come quite to look forward to having a whole week at home. . . . We are still fighting in Germany, though there is some chance of a settlement. Among prospective journeys I have one to Sweden and another to

America. . . . The last time I was in Germany I was in
the Siegen country. I thought, as I passed through by
rail, it was the best scenery in Germany, bar the Bavarian
Highlands. . . . Yours always,

'S. G. T. '

'P.S.—I have now a pile of some thirty letters to
answer. I ought to answer half before going to bed.'

'3 Queen's Road Villas, Battersea: April 14, 1880.

'Dear Bess,—I have again been to France, Belgium,
and Germany for a few days, during which I hoped to run
up at least as far as Coblentz, if not to Wiesbaden. I
would not write you till I knew if I could come, but was
called home from Ruhrort, where I had a long and tedious
business, to meet a man from America, and so was prevented
doing so, much to my disappointment. It becomes more
of a drive every week. Everything both abroad and at
home falls on me, and it is enough.

'I am negotiating for a sale of my German rights, so
as to have something in hand. There is also more fighting
to do in Germany and elsewhere. We have had nothing
in papers now, except now and again a paragraph such
as enclosed. The affair is going well; but it is so big that
it requires perpetual attention, and guarding and watching
with practical work. Percy takes most of the practical
supervision at home and I the rest, and all abroad.

'This is an egotistical spin!

'. . . We are trying hard to get rooms in town as
soon as we can. Probably shan't succeed till June. . . .

'S. G. THOMAS.'

Already the inevitable effects of this over-worked
existence were visible, and doubtless deadly disease was
already at work sapping the very citadel of the vital

forces; but he had no suspicion as yet of the need for
care. It can have been no unhappy life that he led; that
which for years had been his supreme object had been
achieved; his remaining anxieties were of no poignant
kind, and ceaseless activity (however it might physically
wear and tear him) was always a keen pleasure to his
eager nature. Meanwhile the process was everywhere
triumphant on the Continent, and at Middlesbrough Mr.
Richards, with the co-operation of Thomas and of Gilchrist,
was still perfecting mechanical details more and more.

At the spring meeting of the Iron and Steel Institute
the basic process was still the main topic of interest (as
it continued to be at many successive meetings), and of
course the meeting brought new cares to Thomas. The
next letter we quote refers to it.

To Miss Burton

'Queen's Road Villas, May 10, 1880.

'. . . I am awfully ungrateful not to have written
before to thank you for your charming letter and delightful and most useful little present. . . . As usual I am
fairly busy. Last week the Iron and Steel Institute meeting, which went off fairly well. I enclose a report.

'I had to be entertaining people every evening, which
was the most fatiguing thing of all to me. I introduced
Lil to a dozen of the leading engineers of the world in
one evening, which amused her considerably.

'There are still many questions open which cause
anxiety and work; but on the whole things going not
amiss. . . .

'I am trying to get things in order, so that I may go
to America in the autumn if possible. . . .'

In constant journeying to and fro, the summer of 1880

wore away, until the time came for the meeting of the
Iron and Steel Institute at Düsseldorf in August. Thomas
took his sister to Miss Burton in Wiesbaden early in that
month, and the former attended the meeting with him.

Thomas writes from Wiesbaden to his mother on
August 8 :—

'Got here at seven yesterday. A gorgeous reception
from B., who looks well. Stopped in, chatting, all evening.
I sleep at the best hotel. We are now going to Wood ;
shall be here to-morrow night. All very kind and nice.'

Dephosphorisation was as usual the leading topic at
Düsseldorf. Sidney's sister sent home the following report
of her brother's speech on the subject :—

'Sidney's speech on dephosphorising. Friends all
round ; room crammed ; perfect quiet. Prof. Turner spoke
first, then Siemens and Wedding—then Sid. Splendidly !
Clear, ringing, metallic utterance—good delivery, to the
point, *i.e.*, cost and general results. No nervousness
perceptible to the outer world (Mr. Justice[1] was the only
one besides myself who saw he *was* nervous ; shows he
knows him well). I was frightfully nervous for him at
first, but soon I found I had no need to be. He was the
only speaker during the whole week's meetings who was
clapped on standing, and he was so clapped warmly, and also
interrupted for applause. President Ed. Williams requested
him to stop on the platform to be questioned, and many
friends chaffed him afterwards about having struck a theatrical attitude. Then Snelus and Riley spoke and Massenez.[2]
I was quite an impartial witness, prepared to criticise
severely,—as I always do *him* !'

[1] Thomas's Patent Agent—a
personal friend.
[2] Herr Massenez, Director of the
Hoerde Works, was an early and
zealous supporter of the process,
and gave it much help.

During the meeting there was an excursion to the Rhenish Steel Works in Meiderich, where the process was seen in operation. The excitement and interest in the 'blows' were intense. Mr. Richards says:—

'It was most difficult to get near the workmen who were testing the samples, so great was the crush and the desire to obtain a piece of the metal; and the wonder was that the metal was so well blown and so low in phosphorus, considering the circumstances under which the operation was performed.'

The meeting was wound up by an excursion to Cologne and Coblentz, of which Thomas gives brief account to his mother on one of his customary post-cards:—

'Coblentz: August 1880.

'Dearest Mother,—Another awful round of pleasure yesterday. First by train to Bingen, with lunch on the way, 800 of us, about. Then special steamers down to Cologne; lovely weather and lovely scenery everywhere. Lil introduced to thirty or forty new acquaintances. At Coblentz taken through wine cellars, then through Empress's Palace; then a gorgeous dinner. Stopped there too late to go on to Wiesbaden, so remained here. We go on to W. at ten this morning. The meeting a great success. I have been fêted and petted ridiculously. At Essen on September 3.'

A little later he writes (still on a post-card):—

'Bochum : September 5, 1880.

'Here all day yesterday; over Works adjoining, &c. Dinner with the Director; more Works. Wine in evening with three directors; very hot. . . . I am now on way to Hoerde and Magdeburg; at Stassfurt on September 7.'

From Magdeburg he writes to his mother :—

'September 6, 1880.

'Dearest M.,—Here I am again on the move. Now on way to Stassfurt, to see the great Salt Works, which I hope to utilise in phosphate-making. I then go through Dresden (half an hour to see the Picture [3] again) to Wittkowitz. .... I *expect* and hope to call at Wiesbaden about the 12th, but may not be able to stop out so long. Spent yesterday afternoon with Massenez and the H. Y. . . . Mind you have rides with aunt and Miss B. regularly. Love to all.—Yours ever,

'S. G. T.'

From Stassfurt he writes to his sister at Wiesbaden :—

'Tuesday, September 7, 1880.

'Lieber Kleinchen. Hie bin ich angekommen gestern at eight (nicht unterstrichen), habe besucht grosser Fabrik wo vu insisted on mich die thur zu zeigen, bis ich habe developed das ich in solchefalle, it would be my painful duty to obliterate aller spuren von ihren Fabrik wurden.'

Meanwhile his sister had been writing home from Wiesbaden under date of September 4 :—

'Dearest Mother,—Sid arrived yesterday at four; we were at station to meet him. He has a cold and we insist on his staying a day or two to get right. He goes back to Luxembourg and Longwy; will be back on Monday.'

This 'cold,' which was to 'get right' in a 'day or two,'

---

[3] The Sistine Madonna. This was such a favourite with Thomas that he always made time, when- ever he was near Dresden, for a pilgrimage to it.

has a mournful and knell-like sound to us who know the end, and the short but sharp attack which he was nursed through in Wiesbaden was a matter of serious anxiety to his sister and cousin. As yet, however, he persisted there was nothing seriously wrong with him, and the wearying journeying to and fro was continued throughout this year.

The fatigue involved will be sufficiently obvious—a fatigue especially dangerous in the severe winters of 1879–80 and 1880–81.

Early in 1881, however, it began to be clear that such voyaging (with all the necessarily concomitant changes of temperature) must, at any rate in winter time, be discontinued. The cough persisted, and his uncle, Dr. Burnie of Bradford, whom he consulted, detected grave lung mischief. Even London fogs must be avoided. Thomas was persuaded with difficulty to go for a time with his sister to the Isle of Wight, and to take for a brief period such rest as his enormous correspondence would allow him.

The following letters belong to this time :—

To his Mother

'Esplanade, Ventnor : February 1881.

'Dearest Mother,—Two bedrooms, large and facing south and sea, and a ditto ditto sitting-room. Bright, sunny, but cold here. Thermo. outside, at three to-day, 45°; yesterday 50°. Am really much better—cough only very little in evening. Been out all day. Lil as good as can be. I fear we shan't be able to quarrel; she looks after, pets, bullies, worries and amuses me to perfection. You have nothing at all to bother about as regards your robustious children. Hotel slow; though good of its kind. Ventnor prettyish. Love to all. Look after yourself.—Yours ever lovingly.'

To Mrs. Burnie

'Marine Villa, Esplanade, Ventnor:
March 1, 1881.

'My dear Aunt,—It is very kind of you all to trouble about me and my small ailments. I am certainly the better for coming here—decidedly so; though still weak as to breathing arrangements. The weather here is bright and fine, and sunny most days. Some days have been exquisitely bright and blue-skied. It is, however, dull enough, as I can only walk to a limited extent, and there are too many hills to make riding very attractive. Lil has got a girl with her who amuses her much, and me somewhat. My "Bricks without Straw" was bought. I fancy Trübner publishes here. I am in hopes of seeing you in March, that is, if I am able to get North—as I expect to, about the 15th, for a meeting. I rather chafe at being so absolutely tied up just now, when there is plenty to do elsewhere; but it might be worse. Lil is, I think, enjoying herself as she does generally, and is certainly wonderfully well; she is a bright little companion. Your friends have done well to go to Grange. It is a very pretty place in itself, and within reach of still prettier. Please tell my uncle I am following his advice as nearly as may be in all things. With best love to all,

'Yours ever affectionately,

'S. G. THOMAS.'

L

CHAPTER XIII

A VISIT TO THE UNITED STATES

HOWEVER, he did not go to Yorkshire on March 15, as he seems here to have contemplated; for by that time he was on the Atlantic. Circumstances induced him suddenly to determine upon a visit to the United States, with a view to defence of the patent position there. He sailed for New York in the 'Marathon' on March 11, 1881. He was received with open arms by the worlds of iron and steel and applied science. The following letters have been preserved :—

To his Mother

'March 26, 1881.

'Dearest M.,—Got into New York at seven Thursday evening. Laureau came on board to ask me to stop at Holley's. Went with him to concert and to see Broadway. At concert met the Swede Lilienberg. Next morning Maynard came on board and we went to Holley's. Made a lot of calls ; saw chief buildings ; travelled four times on elevated railroad. Was introduced to about twenty people ; dined with Holley at a Palace, far and away above our Criterion. Evening dined at Hewitt's, late Mayor of New York, and with Cooper, the founder of Cooper Institute, a bright, intelligent, and active old boy of ninety-two, who has donated about $2,000,000 to public purposes, and now educates in highest branches 1,800 folks yearly. Absolutely in evening went to Opera with Miss Hewitt and her father. 'Favorita.' Fine house, but overpowering amount of talking. From Opera to Century Club. Am already a member of three great clubs, with free access to Society of Engineers &c. Have invitations for summer to Lake Champlain, North Jersey, and the Lord knows where besides. I am to be dined by forty men next week, alas ! If I don't get spoilt, I shall be surprised. New York is a quarter of a century ahead of London, (1) in telegraph facilities, (2) in buildings, (3) in elevated railways and tram cars, (4) in size and convenient arrangements, (5) decoration of houses, (6) in small conveniences.

'Monday evening called on Carnegie and others. Lunched at Delmonico's. Introduced to more people. Dined at University Club with Holley ; beautifully decorated. I find they are tremendously ahead of us in decoration. After to Brooklyn Club.

' Sunday, went to Beecher's with Mrs. H. and Miss G. Plymouth Church hideous, but crammed. B. preached for one and a quarter hour—most eloquent, original, and sometimes *outré* sermon or address. He is obviously a man of immense power. Parted with regret from Mrs. H. and Miss A., and to Maynard's—pleasant afternoon. Called on Raymond, a very clever fellow, who is engineer, poet, novelist, editor, man of business, musician, composer, and Sunday school teacher, all at the same time.

'I want to get out of New York as soon as possible. I only regret not having you both. I don't like American girls so far—bar some. They have vivacity and dash enough to set up a city, and have a good time, in other words, have their own way, undoubtedly.

'March 31, 1881.

'Yet more dissipation and enjoyment; calls, dinners, opera with Carnegie, &c. Tuesday evening, up Hudson to Albany in palatial steamer with Holley. Arrived Albany 6 A.M.; over State Capitol, an enormous building still in progress. Senate and Representative Chambers superb; some of architecture finest I have ever seen; decoration massive and grand, in excellent taste.

'Charming dinner in fine old-fashioned house of proprietor of Works. Mrs. C. very pleasant and lively. Next morning I wanted to go to Works; but Mr. C. insisted on driving me to his country house, and showing us some miles of hothouses with wonderful varieties of plants from every quarter of globe; plants worth some 20,000. Collection of 13,000 butterflies.

'Back to New York by train down Hudson Valley, which is very lovely,—more so than the Rhine on whole.

'That confounded dinner comes off to-morrow. Continue brilliant; but love hard work,—not to be over-dined.

'The people have to a stranger few deficiencies, except a too evident money-worship, and (whence the money-worship proceeds) a reckless way of spending. They are hospitality itself.'

The next letter is written in the margin of a copy of the 'Iron Age':—

'Saturday, April 2.

'Dearest Mother,—The dinner is happily past and I actually enjoyed it, partly. It was dreadful sitting for three hours and being bepraised; but the speakers were really clever and witty in the extreme—alternating between flights of real eloquence and the most fanciful word-fun and wildest jokes. The actual dinner was, of course, superb, costing about 200*l.*

'I trust the sale is practically settled. I go to Philadelphia to-day about it; then back here for a day or two; then to Bethlehem, &c. I am invited to about twenty dinners, and to stop at about a score of houses all over the States—passes on lines where I *don't* want to go, &c. &c. Of course it is evanescent, but amusing.

'I got through my speech fairly, I think. I had brought over a first-class one, but couldn't think of a bit, so started on quite another line . . .—Yours ever affectionately,

'SIDNEY.

After this comes a sort of post-card and letter diary to his sister and mother, which we partly reproduce :—

'April 7, 1881.

'Back from delightful two days at Bethlehem. Boundless hospitality; enjoyed and benefited by it much. Fritz is a charming fellow. Go to theatre to-night. Bessemer matter still hangs.

'April 8.

'Was last night at theatre : saw Daymond in "Fush;" a wild comedy, great fun, but absurd. This morning, long interview with lawyers; shan't get business done for three weeks at least. Then more interviews. Then Dr. Raymond drove me through Brooklyn's beautiful park to Coney Island, eight miles away; beautiful spin behind splendid horses. Am now at theatre, having been at Cooper Institute. All right.

'Chattanooga, April 14.

'Left Washington yesterday at 7 A.M. ; travelled there by palace car, got here at 8 A.M. Country very picturesque, but very few substantial houses, wood shanties being the bulk. Travelled very comfortably; went to bed regularly at night, eating copiously by way. All country, but

indeed mountainous. Shall be here three or four days.
Constantly thinking of you. I came south now, which I
ought not to have done, to keep promises to you and avoid
cold of the north. Have had long morning's interesting
and instructive interviewing. A beautiful country and
lovely day; feel quite brilliant. Am thinking of settling
in the U.S. if the mother and you will come.'

'Grand Hotel, Chicago : April 20.

'Dearest Mother,—Wrote you yesterday from Cincin-
nati. Came over here by sleeping car very comfortably.
Been all day at the Works; of course well received.
To-morrow more Works; then to Joliet [Works], Springfield
[Works], and back to Pittsburgh, and so to New York.
Weather here coldish, but bright mornings. Some snow
left in streets and lake frozen. Chicago certainly is a
marvel; one can't credit it with being a fifty years old
town, and a ten years old phœnix. It looks enormously
prosperous and substantial; the country flat and uninteresting enough. I take to palace-car sleeping travelling
greatly; it truly makes distance no object, except to the
purse. Constantly thinking of you; sometimes somewhat
home-folk-sick.—Yours,

'S. G. T.'

'April 21.

'Got yours and L.'s. Please always say specifically if
you are well. I have written by every mail since I have
been away. Two Works to-day. About Chicago a mass
of fine residence houses, and also as fine business places.
Forsythe very kind. Hotel 500 rooms, good specimen of
caravanserai; ground floor, railway bureau, barbierstube,
assurance office, electric baths, &c. &c. Bonne cuisine, but
not much real comfort.'

'Fort Wayne, Illinois : April 23.

' Joliet Works very interesting. Well received. Splendid day there; dining and supping with manager. Arrived at Springfield, Illinois, at 10 A.M. All day at Works. Has pleasant houses, a gorgeous State Capitol, and streets in which the mud, without exaggeration, two feet in thickness; fine houses and shops of brick and stone are jostled by wood cabins in the most curious way. Works very interesting to me, as they are working Pernot process, in which I am greatly interested.'

' The pleasant little wife of —— explained to me the social points of Springfield thus. She belongs to a French class, an Elocution class, a Shakespeare class, an Art Club and a Married Folks Club. I find all the married women here go to classes for languages, or literature, or something.'

Next comes a regular birthday letter to his mother :—

'Pittsburgh, April 28, 1881.

' Dearest Mother,—I calculate this should reach you on your birthday. I only wish I could be with you too, or you with me. I shall be thinking of you, then, specially, and hoping you may be bright and well, and as happy as the best little mother in the world should be. What Carlyle says of his wife I often feel of you, especially with an ocean between us—that I never can or shall appreciate one tenth of what you have been to us all. Now to my usual egotistical chronicle. I left myself on Monday morning, when, after a pleasant call I adjourned to the great Edgar Thompson's Steel Works of Carnegie's. Spent there many hours with advantage; dining with manager—a vigorous and singularly able man. Home to hotel and business till bedtime. Tuesday morning, was joined by Holley from New York; spent all day driving about to Works. Admirably received, of course. Much interesting, and all shown without reserve or hesitation. In evening went to theatre with Andy, Carnegie's brother; much amused. Early next morning to Edgar Thompson again, with Holley; then to another Bess. Works, a party being made up to accompany us. In evening (last evening), a dinner by a dozen or so of leading iron men to S. G. T. Brilliant dinner; then Chairman called on every individual to make a speech. I—poor I—was lugged in by every speaker, of course. I had to orate twice, which I did with commendable brevity. It is a dreadful nuisance, this being talked at, and expected to talk, and what is worse, be funny. The American does, however, manage to let off a wonderful lot of clever and humorous things. By practice, I think I should learn to grind out a good thing once a month or so. We got home at 1 A.M. and left Pittsburgh at 8 by palace car to Johnstown, through beautiful scenery, along the Pennsylvanian Road. At Johnstown, one objective was the Cambrian Works, an enormous and most flourishing concern. Met there a lot of people dining and supping with the manager. Came by sleeping car to New York (29th) morning. The Bess. people have paid money; but I haven't yet received it. *6 P.M.*—Yet another pause : at last I have received a good bit, at least, of the Bess. money.

'I shall now be in New York some days. I give a lecture at School of Mines on Tuesday. Century Club to-morrow. Go to Worcester on Wednesday. Ever so much love, dearest mother, and ever so many happy returns of the day.—Yours ever affectionately.'

To his Mother

'Near Buffalo, on Erie Railway : May 7, 1881.

'Dearest Mother,—In last I was starting for Hartford. I met L. at station, and waited till last moment for the entertaining young person I had looked to escorting down ; but alas ! she appeared not,—so like a philosopher I consoled myself with " Well, it's just as well not ! " We had a very pleasant journey down, as we found the leading American landscape painter in the cars and L. introduced us. We talked no end. He [is] just back from Mexico—has been all over Europe, Greece, Turkey, &c.—and much in South America ; very pleasant. H. met us at station—drove to hotel. He had wired his daughter not to come, as no ladies were there. Went to meeting of Mechanical Engineers ; then to State Capitol. Such a magnificent place, in a Moorish-Venetian style— all in white marble outside, with much coloured marble inside, the staircases and panellings massive carved marble —altogether as nearly perfect as an architectural thing can be, on a little hill, laid out as a park with river running nearly round it. If it were in Italy, you would have said " Now, there's a thing you modern architects could never do, nor any of your men of the Steel Age," and troops of pilgrims would go to see it. In evening a banquet,—I located between President and ex-Governor H. Had to respond to Iron and Steel Institute, and spoke very badly, after which three cheers for S. G. T. No, I am not spoiled ; I take it just for what it is worth. A number of brilliant witty speeches and two *worse* than mine, " to my great content," as Pepys has it. Next day visited Works : very interesting. I was only introduced to sixty people at Hartford : asked to stop, but declined, and came back to New York, and on direct to Niagara. I am now on way there, of course in a Pullman. Am always well. Weather bright.—Ever yours.'

Next follows a series of post-cards :—

'May 9.

'On train from Niagara. The big Falls are certainly
well worth seeing. I arrived at Niagara at noon yesterday : drove [over] a light and graceful suspension bridge
to hotel on Canada side. At dinner picked up amusing young Englishman, fresh from a visit to Texas. We
spent afternoon under and over the Falls, which I won't
attempt to describe.

'In evening arrived a young London banker, known
to ——. We had a pleasant trio talk and a glorious view
of Falls by moonlight. This morning viewed them again
from all possible points, to my great pleasure, and finally
I go at 2 P.M. for New York. I start Tuesday morning
for P. R. run on Pennsylvania.'

'Belleforte : May 13, 1881.

'Dearest Mother,—Have really had a good time for
last three days. On Monday, as I wrote, I stopped at
Windsor. Tuesday, 9 A.M. I, Carnegie, and a Dr. Gilchrist
started for a place near Tyrone on the Pennsylvanian Railroad. We arrived at 9 P.M., having picked up on our way
a special car, with a railway man and two Pittsburgh
partners of Andrew Carnegie's. In this car we have slept
for three nights, and fed gorgeously. Real fun. Gorgeous
scenery, beautiful mines, grand furnaces, and lots of new
people. Had several long drives, and saw no end of the
interior country. It has really done me good. I now
go back again, three hundred miles or so, to Philadelphia,
Washington, &c. The woods are delicious in their first
greens. I am always longing for you two folk, which
spoils my enjoyment and makes me look forward to June 8.'

'Harrisburg : May 15.

'Dearest Mother,—Slept at Altona, pretty place chiefly
remarkable for containing all the works of the Great

Pennsylvanian Railroad. Carnegie introduced me to some engineers, and at 7 A.M. was round the "shops," that is engineering; then came down with Dr. Dudley here, a four hours' ride through beautiful scenery. Afternoon at very interesting Steel Works; slept, and now off to Philadelphia.'

'Philadelphia: May 15.

'Rather tired of hotel life, with its monotony and numerous dishes. Have been so busy that I have presented no introductions; only seen young Conway and Mr. Holland, a friend of Aunt A.'s. Carnegie and his party sail on June 1. Want me to go with them: am afraid I shall be unable to sail till 25th.'

'New York.

'As I post-carded, was at theatre last night. Enjoyed it much; that is, it made me quite miserable. A melodrama, remarkably well acted and written. Working hard all day; am tired; been only introduced to six people to-day. All the men I meet are *the* most remarkable in America, are also "gorgeous," "lovely," "princely," "magnificent," "superb," "heroic," &c. Have not been introduced to an ordinary mortal yet.'

These epistles from the United States seem to us to give a real picture of Thomas, with all his eager energy, vivid sensibility, and keen delight in life and its spectacles. It will be perceived that he took as much genuine pleasure in architecture or scenery as in converters and smelting furnaces. He was still the same Sidney Thomas who knew his Dulwich Gallery by heart, just as he was still the same Sidney Thomas who had stood for hours in Grove Lane watching the construction of the main sewers.

He hardly, however, gives us an adequate idea of

the reception accorded to him by the hospitable American ironmasters and scientists. His ingenuous modesty leads him constantly to understate the interest that was exhibited in the solver of the dephosphorisation question.

He has given us his own impressions of the country and people. The following extracts from the New York 'Iron Age'[1] give something of the impression he produced.

'On Thursday last Mr. Sidney Gilchrist Thomas, whose name is now so familiar to every one even remotely connected with the iron and steel industries of the world, arrived in this city from England.

'The hospitality upon which Americans justly pride themselves, and a desire to tender Mr. Thomas the courtesies to which his genius and achievements entitle him, will undoubtedly assure him a reception worthy alike the hosts and the guest. His youth, . . . his modest bearing and unassuming manners, will gain for him many strong personal friends. Though appearing to be rather a scholar than a man of business, his familiarity with the practical details of his profession and enlightened and broad views of matters pertaining to the trade rapidly efface the first impression. More perhaps than any other man now living, Mr. Thomas represents a class of inventors to whom the future belongs, and his success is a striking instance of the correctness of the principles which have guided his work. His efforts will be an encouragement to those who seek for improvements of present appliances and processes by the slow and laborious method of studying the causes which arrest further progress and devising means for their removal. In the popular mind an invention is little more than a lucky idea, which, if it happen to hit the right

[1] Of March 31, 1881, and February 26, 1885.

thing at the right time, brings wealth and glory to the
one who has been favoured with the inspiration, and there
is a large class of men who do little more than hold themselves in readiness for such fortunate accidents. Mr.
Thomas does not belong to this class.'

In a long memoir published four years later, after
Sidney's untimely death, the same paper gives a sketch
of his personal appearance as it struck his American
friends :—

'Mr. Thomas will be well remembered in this country.
His personal appearance was striking and peculiar. He
received honours and awards modestly, and his boyish face,
careless dress, and exaggerated forehead strongly suggested struggling genius rather than world-renowned
success. He was . . . always companionable, bright and
entertaining. Those who knew him felt for him a strong
attachment.'

Mr. Carnegie of Pittsburgh (the author of 'Triumphant
Democracy'), who is so often spoken of in the foregoing
letters, says of Thomas :—

'The first thought that passed through my mind when
I saw him was, "He's a genius." I never saw one who
so completely separated in himself talent from that
indescribable thing we call *genius*. I cannot think anyone
would use the words "able" or "talented" in connection with him. All about him seemed extraordinary.
Appearance, manner, dress, voice, gesture, all said without
saying, "Listen to me, attend! I am not of the routine
world, I walk no beaten track ; from the unexplored and
unknown I bring you fruit." He did not need to speak
this ; his manner and gaze made you see and feel it. He
had only to appear and we bowed before his power. I have

never met a man who carried me so completely away as
Sidney Thomas did.'

Mr. Carnegie has also described Thomas (the 'pale
Gladstonian-looking youth' as he calls him) in his 'An
American Four-in-Hand in Britain.'[2]

2 At pp. 85-90.

CHAPTER XIV

HEALTH FAILS IN EARNEST

EARLY in July of 1881, Thomas, having accomplished his purposes across the Atlantic, returned to England. Upon his return he prepared, in conjunction with Mr. Gilchrist, a 'Note on Current Dephosphorising Practice,' for the autumn meeting of the Iron and Steel Institute. This 'Note' is mainly concerned with technical details, but furnishes some interesting statistics of the progress already achieved, just three years, as it then was, after the famous Paris meeting when the original papers of the two cousins had been passed over, and but little more than two years since working on a large scale had been begun.

'The present current manufacture of dephosphorised steel amounts,' said the young authors, 'to between 27,000 and 29,000 tons a month. It may be added that the make for November, and probably for October, will considerably exceed 30,000 tons, or say at the rate of 360,000 tons a year,[1] while, in the course of the next few months, twelve more Converters, now nearly finished, will come into operation, bringing the yearly make up to considerably over half a million tons.

.   .   .   .

'As to the quality of the steel produced, the rapid

---

[1] In our final chapter we give some particulars of the growth of basic steel-making and the *present* rate of production. The figures had been even two years after this paper more than doubled. *Sed cf.* 'Conclusion,' *post.*

extension of its employment for every purpose for which
Bessemer steel has ever been used (excepting perhaps the
manufacture of Bessemer tool steel) is the best evidence.
That dephosphorised steel is even superior to hematite steel
for certain purposes, such as rails and other plates and
wire, is now pretty well agreed. The total number of
Converters at present regularly working on phosphoric
iron is thirty-six, of which, however, eight or nine are less
than four tons capacity. Thirty more Converters, specially
designed for the process, are now under construction.
Several Siemens furnaces have been in regular work for
some time, but details of their operations must be reserved
for the present.'

Thomas spent a portion of August with his mother and
sister at Sandgate; but soon betook himself to renewed
continental journeyings. The following correspondence is
of this period:—

To his Mother

'Vienna: September 12, 1881, 8 A.M.

' Came here last afternoon. Went to theatre, and bed,
after walking about a magnificent city, all bright. People
pleasant looking. The Kupelwieser charming to last moment; hospitality almost too great. Kupelwieser wanted
to come to Vienna with me, in order that he might show
me about. Lil to go there next year if she behave. They
will probably visit us in spring for a day or two. Shall go
to-night to Wiesbaden, then Luxembourg, Longwy, to
Bonn, where I shall be very glad to get. Frankfort, September 13.'

'Metz: September 18, 1881.

'Dearest Mother,—Confirm mine of this morning.
Got here at 5 P.M. Do some Works; on in morning I
expect, for a few hours to Wiesbaden; then down to

Dortmund on Sunday night or Monday. Again a lovely day, just hot enough and very bright. Holley not very brilliant. I all right.'

'Biebrich: September 19, 1881, Sunday, 5 P.M.

'Got to Neuen Kirchen at 5 last night. Deputation to meet us at station; did Works. Dined with owners, then beer and wine with all engineers till 12. Up at 6 A.M., off to K. Saw W. off at noon on a six hours' rail ride to Wiesbaden.'

'Dortmund : September 20.

'Got here at 11 last night. Spent three hours in Cologne and good time around Dom, which is magnificent. This morning went to H. M. M. J. returns to-morrow night, so I shall stop here till Thursday night, then to Ruhrort on Saturday. We shan't join the Holley party after all.'

Alas! with the returning autumn it became absolutely necessary to suspend activity, if, indeed, Thomas could ever be said to suspend activity. We all of us remember the story of the man who was placed in a chamber from which there was no escape, and the walls and floor and ceiling of which very slowly, but very surely, contracted and drew together. In such a chamber was the bright young life now, as it were, imprisoned. Manifestly this winter could not be spent as former ones had been; for the lung trouble grew worse rather than better. Thomas was strongly pressed both by medical and lay advisers to spend the dangerous months in the south of France. There is before us a letter from Mr. Lushington in which this view is forcibly urged. He writes in November :—

'I am very sorry to hear that you have got out of health and are recommended to go south for the winter. I hope you will lose no time in complying with the recommendation, and get out of this climate and through France

M

before the winter sets in, even if it is only a matter of extra precaution. The old French proverb, *la lame use le fourreau*, is not one which is safe to neglect; and it would be very foolish in you to overtax the sheath of your intellectual identity by hard work, in despite of any temporary weakness just now. I trust you have every right to anticipate a long and prosperous career as the reward of your scientific labours; but the chances of health are not things to trifle with. I am sure you will not resent, and I hope you will not be tempted to disregard, the advice.'

However, Thomas could not see his way to leaving England at this time, and he compromised matters by wintering at Torquay, whence he returned to London in the spring of 1882,—only to find that he must, until summer finally set in, betake himself to Ventnor.

At Torquay he had both his mother and sister with him, as well as many visitors. His sister writes concerning this period :—

' I remember much work—incessant writing—a great deal of fun and merriment. A favourite game with us was anagram making. A novel read at this time, and much appreciated, was Mrs. Burnett's "That Lass o' Lowrie's." Sidney was always ready to turn everything into a joke, including his own " petty ailments," as he insisted on calling them. One of these " petty ailments " was an inability to walk fast or far, which was just beginning to show itself. Alas! Sidney had until then been a vigorous walker indeed, both as to pace and distance.

' I had some delightful rides with him. Every movement of his horse was the text for comical disquisitions on what was passing in the quadruped's mind, and speculations on the animal's view of life in general, and of his rider in particular.

'We got up the history of Devonshire, and made expeditions to various points which our studies led us to imagine would be interesting. Once or twice we rode in the Brixham direction, and on such occasions we would return home with pockets weighed down with ironstone.'

From Torquay Sidney wrote to Wiesbaden :-

To Miss Burton

'My dear Bess,—Torquay is *very* slow. *That* is the predominating idea with me at present. I must confess it is pretty enough, and that it is sunnier (when there is sun) than elsewhere ; but I am not constructed to revel in polite watering-places.

' Lil and the Mother very good, and insist on amusing me.

' I am well enough on the whole,—decidedly better. get all my letters here, so can keep things going. There is plenty to do.

' We have drives frequently, and hope some day to get out to Dartmoor.

' I may go over to America again in the spring. In the summer (or rather September) all the Iron and Steel Institute go over to Vienna. I expect I shall take Lil and have a " good time ; " perhaps going to Italy as well.'

At Ventnor Thomas was with his sister. The latter says :—

' We led much the same life at Ventnor as at Torquay ; more rides, more reading, more work, more fun. It would have been very pleasant had not the days been darkened for me by increasing anxiety concerning him. I remember Sidney spending a whole morning on the sands with myself and a girl friend constructing a dam, aqueduct, and embanked canal, diverting the course of a little rivulet.'

Thomas and Mr. Gilchrist were meantime engaged in preparing a paper on ' The Manufacture of Steel and Ingot Iron from Phosphoric Pig Iron,' which was read to the Society of Arts in April 1882, and received the Society's medal. It gives so clear an account of the whole matter that we cannot resist making some rather copious extracts from it.

'Lord Palmerston's terse and accurate definition of dirt,' the authors begin, ' as " matter in the wrong place," may with singular appropriateness be applied to the phosphorus which, while itself a substance of considerable commercial value, is unfortunately so generally associated with iron ores to the great detriment of their utility. . . .

' Dephosphorisation endeavours to relegate this wrongly placed matter, if not into its right place, at least into a neutral position, where it can do no active mischief. The actual importance and scope of dephosphorisation in its application to steel-making is most readily realised if we bear in mind :—

' 1. That on a rough estimate about nine-tenths of the whole deposits of iron ore in Europe contain more than one part of phosphorus for every thousand parts of iron.

' 2. That in the smelting of iron ore in the blast furnace to form pig iron (the first step in the conversion of iron ore to a malleable material), no phosphorus is removed, so that, practically, all the phosphorus found in the ore is found also in the pig.

' 3. That in neither of the two great steel-making processes, as ordinarily carried out, is phosphorus removed, so that all the phosphorus found in the pig is, under ordinary circumstances, found also in the steel into which it is converted.

' That the presence of more than one part of phosphorus in a thousand of steel is not permissible where reliable quality is necessary, phosphorus, as is well known, causing in steel extreme brittleness at ordinary temperatures.

'. . . The non-phosphoric ores are confined in England to Cumberland, Lancashire, the Forest of Dean, and two or three other very limited areas, as Weardale, Mwndy. . . .

' On the other hand, the whole of the ores of Scotland, Yorkshire—including the vast deposits of Cleveland with its yearly output of 6,500,000 tons—North and South Wales, Shropshire, and Staffordshire, and the great belt of country extending from Wiltshire across Oxfordshire and Northamptonshire to Lincolnshire, are phosphoric. These deposits are of so enormous an extent as to render it very difficult to calculate their probable content of ironstone ; but an attentive examination of their area justifies the conclusion that the nonphosphoric ores are in Great Britain at least ten times more abundant than the purer kinds. . . .

' On the Continent also all the largest deposits, with the exception of those of Spain and Sweden, are phosphoric. The great phosphoric ironstone region shared between Luxembourg, the Meurthe-et-Moselle, Alsace-Lorraine, and Belgium, is alone more considerable than all the other deposits of Northern Europe together.

' In America the deposits of Bessemer ore are very large, but are greatly exceeded in magnitude by the great phosphoric ore-tracts of Pennsylvania, Alabama, Tennessee, and Virginia, and it is highly probable that the centre of the steel manufacture of the United States will on this account gradually gravitate southwards. . . .[2]

' How does it happen that there are 9,000,000 tons of pig-iron annually turned into the unquestionably inferior material known as puddled iron, while only 5,500,000 tons

[2] This rediction seems now in course of fulfilment.

are converted into the superior form of steel or ingot iron—
particularly when so great an economy of fuel and labour
could have been effected by turning the whole into the higher
class materials? May it not be fairly contended that it is
really nothing but the absence of a practical and economical system of dephosphorisation that could justify the
existence of such an anomaly? It is now proposed to
show that there is no reason for its continuance; since the
development and modifications introduced during the last
four years enable steel of any desired purity, as regards
freedom not only from phosphorus but from silicon and
sulphur, to be produced readily and economically from the
most highly phosphoric kinds of pig iron.

'The Bessemer process with concurrent dephosphorisa-
tion—as now practised at the Middlesbro' Works of Bolckow,
Vaughan, & Co. (who, under the able guidance of Mr.
Windsor Richards, have been the pioneers of the new
industry) and thirteen other Works in France, Belgium,
Germany, Austria, and Russia—is carried out as follows:—

'The Bessemer vessel is lined with magnesian lime,
which has been previously subjected to an intense white
heat, and so brought to a condition of density, tenacity,
and hardness as far as possible removed from the conditions of the material generally known as "well-burnt
lime," and more closely resembling granite or flint. This
material, which for brevity is known as "shrunk lime"
(as in course of preparation it shrinks to one half the
bulk of ordinary lime), is used either in the form of bricks
or in admixture with tar, as a rammed or "slurry" lining,
this being substituted for the ordinary silica brick or
silicious ganister [3] lining of the hematite process.

[3] For the meaning of 'ganister'
see *ante*, p.32. This whole description should be carefully compared
with Bessemer's account of the
original process there given.

'Before the metal, which may be either employed direct
from the blast furnace without intervening re-melting, or,
if for any reason this is not convenient, may have been re-
melted in a cupola, is run into the converter, from 15 to
18 per cent. of common "well-burnt" lime is thrown into
the vessel. The metal is then introduced and the charge
is "blown" in the ordinary way to the point at which the
ordinary Bessemer operation is stopped—that is till the
disappearance of the carbon, as indicated by the drop of
the flame. The dephosphorising process requires, however,
to be continued for a further 100 to 300 seconds, this period
of so-called "after blow," which would be prejudicial both
to quality and yield in the ordinary process, being with
phosphoric iron (under conditions permitting of the
removal of phosphorus) that in which the great bulk of the
phosphorus, down indeed to its last traces, is removed.
The termination of the operation is shown by a peculiar
change in the flame and checked by a sample of the metal
being rapidly taken from the turned-down converter,
flattened under the hammer, quenched, and broken, so as
to indicate by its fracture whether the purification is complete. A practised eye can immediately tell whether or
no this is the case. If the metal require further purification, this is effected by a few seconds' further blowing.

'The operation is thus, as will be seen, but little
different from the ordinary Bessemer process. The differ-
ences that have been indicated, viz., the lime lining, the
lime addition, and the after-blow are, however, sufficient
not only to enable the whole of the phosphorus (which
would be otherwise untouched) to be completely removed,
but the silicon, of which inconvenient and even dangerous
quantities are occasionally left in the regular Bessemer
process, is also entirely eliminated, while at least 60 per
cent. of any sulphur (also untouched in the ordinary pro-

cess) which may have been present in the pig is also
expelled.

'It is found, too, that the once dreaded phosphorus is of
most substantial assistance in securing by its combustion
the intense heat necessary for obtaining a successful blow
and hot metal.

'If it is desired to produce "ingot iron," or a metal
differing only from puddled iron by its homogeneity and
solidity, the usual addition of spiegel is omitted, or re-
placed by a half per cent. of rich ferromanganese. . . .
The phosphorus is oxidised by the blast, forming phosphoric acid, which, finding itself in presence of two strong
bases, oxide of iron and lime, unites with the latter of
them to form phosphate of lime, which passes into the
slag. Whether or no there is a transitory formation of
phosphate, making oxide of iron perform the function of
a carrier, is a matter (though interesting theoretically)
which it is needless here to discuss. . . .

'The basic Siemens and Siemens-Martin processes [4]
are carried out upon the same lines as the Bessemer process. The dephosphorisation is very complete, but the
operation takes about five per. cent longer than when pure
material is used: the proportion of lime required is less
than in the Bessemer process, and the wear of the basic
hearth, with suitable arrangements, is not excessive.'

The authors then proceed to discuss questions of cost,
and show the gain by using phosphoric ores (so much
cheaper than hematite ones) in the Bessemer process. 'As
compared with puddling we find that the basic Bessemer
process is more economical in every item except that of

[4] It has been thought unnecessary to describe these processes in
this Memoir. The Bessemer pro-
cess, if thoroughly understood, will suffice to make clear the
utility and importance of the
Thomas-Gilchrist operation. *Ante*
p. 33.

loss of metal and waste of lining—the economy in labour
and fuel being especially notable.'

The whole paper is so logical in its arrangement and so
interesting in its matter that we wish we could reproduce
it in its entirety.

Sidney, as may be seen from what his sister has said
above, although he might tarry in Devon or the Isle of
Wight, could not be induced to rest. The mass of
correspondence and business which his patents in various
countries and other matters connected with his great discovery brought to him was huge indeed, yet he was ever
seeking new avenues of activity. From the first days of
his success he had given with the most generous liberality
to such objects as commended themselves to him. It may
be, as some of us think, that no good can be done to the community by any charity, however enlightened, so long as
the present system of society endures; but at any rate
good may be done to individuals and (in any case) one
cannot help loving the cheerful self-sacrificing giver, who
gives from the abundance of his heart, or because he
honestly believes that he is redressing social injustice, and
not as one merely paying 'ransom' for his riches.

'I would urge him to rest,' says his mother, 'and tell
him that he had done enough for many years at all events:
but the answer to me always was, "You see, mother, I
must, if I live—show that I can work at other things
besides dephosphorisation. Besides I must make more
money still; I have really given so much away that we
shall be hampered in our plans for colonisation, workers'
dwellings, and what not, if I don't!"'

'If I live' is the phrase as quoted by his mother.
Already, it would seem, the thought that it might be that
he would not live, was shaping itself in his mind. He
writes to Mr. Chaloner about this time:—

'27, Tedworth Square, Chelsea: June 20, 1882.

'Dear Chaloner,—I should not trouble about these details, but am fixing up everything so that, in case of accidents, my affairs would stand on a simple and business-like footing for my representatives.

'I wish you *would* look in when near. I am really so tied up that I can't make any calls (though I am obliged to travel a bit from time to time). I should like a chat. I shall probably go to Germany on a Works round in July, and in October go away for six months, I expect.—Yours ever,

'S. G. THOMAS.'

Thomas spent August of this year of 1882 in Guernsey and Jersey with his sister Lilian.

'In Guernsey and Jersey,' she says, 'we spent the happiest month, a month of continual sunshine. We drove almost over the lovely islands, housekeeping merrily together. In Guernsey, Sidney always came with me to the fruit market—a delightful mass of lovely colour. We led a bright, simple life, full of work and fun, fresh air and sunlight.'

The following letters belong to this period:—

To his Mother

'Dearest Mother,—We are wonderfully favoured with bright weather; though coldish wind last two days. We lead such an idle life as ought to shame one; but I manage to keep a little business moving along. I wish you could be here; but at same time feel it doubtful if it is not too much of a journey. The place is dull enough and to spare. Wish someone would come down, but can't recommend anyone to do so. I am not quite sure that it suits me. By the way, I am quite clear the *east* coast would not.

L. the best little woman possible, thoughtful and good to a degree. Have enjoyed my " Middlemarch," which is inimitable, and also some " Nineteenth Century " and other mags. The bound volume " XIXth " for first half '82 is full of interest. I think we could move up to some place where you could come next week: will give this a few more days' trial.

'Am ever so much better, but the east monotony makes me feel a bit cranky. Lil has, of course, told you more than everything. Truest love.—Ever your son,

'SIDNEY.'

To Miss Burton

'Guernsey: August 19, 1882.

'Dear Bess,—I have had the hope all spring of spending some time at Wiesbaden, on my way to Vienna as before; but (like many other hopes) this is, I suppose, to be disappointed. I have not written you for long, as I thought, if I waited, I might write I was coming to see you, or that I was well enough to be too busy to come. However, my unpleasant lung trouble, so far from departing, seems always tightening its hold; so I came here three weeks ago with Lil, preparatory to going away *somewhere* before October,—to Australia or America, it will probably be, or round the world—as I want to make a fight to get some work done yet. . . . Guernsey has fine coast scenery; though inland it is too highly cultivated to be very picturesque. It is small farming pushed to extremities. . . . I am heartily tired of this, as (when one can only walk a hundred yards at a time) is natural enough. We return on Thursday or Friday, and I probably start for *somewhere* in another fortnight.

'I regret the missing Vienna Iron and Steel Meeting immensely. You must come and see [us] when I return from my long trip. We made the acquaintance of some

very delightful Americans who have been stopping at
Kingston, and had planned spending the winter together
in Italy, after going to Vienna together. However, I
expect the long trip is the wiser one.

'The Process is making fair progress. I am much
annoyed at having to leave it now, when so much remains
to be done, and also before our North Eastern Company
is fairly started at 'work. . . . I have, I hear, just been
elected on the Council of the Iron and Steel Institute;
which is rather a pleasant compliment, as the membership
is usually reserved for much older men and greater swells
than I. Lil very bonny and good; makes a most cheerful
companion in a dull place. . . . You may next expect to
hear from me from the Antipodes or elsewhere.—Yours
always,

'SID. G. THOMAS.'

All the while, however, his life (although he knew it
not), was drawing onward to the end, an end which was
so sad because so early. With his return to London and
with the first breath of autumn he again grew worse, and
it was necessary to once more seek refuge in Devonshire,
until Thomas could put his affairs in order and embark
upon the voyage in search of health, which had now
become imperatively needed. He was unable to be present,
as he had much looked forward to being present, at the
autumn meeting of the Iron and Steel Institute, which
this year was held in Vienna.

At this meeting a very unusual honour was conferred
upon him in his absence. Herr Bœumler obtained leave
to present to him in full meeting, on behalf of the Prague
Ironworks Company, a beautifully wrought casket made
exclusively of pig, 'ingot iron' and steel. He said he had
been deputed by his company to present the casket to

Mr. Sidney Gilchrist Thomas, 'as a mark of their appreciation of his genius, as well as to express, though in a somewhat feeble manner, their gratitude for the benefits conferred upon their district by the basic process. He learnt with deep regret that Mr. Thomas was too ill to be present, but he would place it in the hands of Mr. Gilchrist, who would hand it over to their absent friend and benefactor.'

Meanwhile, Thomas had decided to begin the winter in South Africa, and from thence to push on to Australia (Australia had been an attraction to him all his life), taking possibly India by the way. It was arranged that Mr. Honman, a young medical man for whom he had a great liking, should accompany him. Before starting, he wrote the following farewell letter to his constant correspondent at Wiesbaden :—

To Miss Burton

'Kingswear, Devon : October 11, 1882.

'My dear Bess,— . . . I have been here now ten days, and am all the better. It is a *singularly* pretty place— quite the prettiest I know in England. I sail day after to-morrow at noon in the "Conway Castle" for the Cape or Port Elizabeth ; stop there a few weeks—and then to Sydney, Australia. Such at least is my present idea. I am pretty confident that I shall return recruited. Lil and the Mother down with me, looking after me very closely. I tell them they will have nothing to do when I cease to occupy all their time.

'I am dreadfully busy getting necessary papers, deeds, and letters off. I shall have a whole batch to send off to post from on board.

'This is my excuse for a necessary brevity. Kindest regards to all.—Ever yours,

'S. G. T.'

CHAPTER XV

SOUTH AFRICA

ON October 13, 1882, accordingly, Thomas sailed with Mr. Honman for the Cape. His copious correspondence with his sister and mother during his absence seems to us to be very interesting in itself, and to illustrate his own character, his power of making friends, his clear outlook upon things as they were, his rapid grasp of economic conditions, in a very remarkable fashion.

To his Mother and Sister

'Tuesday, October 17, Afternoon : off Madeira.

' Dearest Mother and Child,—Directly you left the ship on Friday, I felt that there were a hundred things I wanted specially and particularly to say to you both, that I had left unsaid ; but as I fancy we should all have been feeling *bad* and badder the longer we put off parting, it was perhaps best as it was. I was on deck till about five. Saturday was fine again, though it got cold in the afternoon. Sunday also coldish and wet most of the day, so I kept in cabin and smoking room mostly. Yesterday warm (65°-72°) and sunny all day ; I sitting on deck and basking in the air from morning to night. To-day, if possible, still more brilliant and the sea mirror-like almost. Madeira looks lovely as we approach. Now as to myself. Saturday and Sunday I only felt middling, with now and again some

chest pain. Yesterday and to-day I have felt *no* pain ; a
prodigious appetite, and generally in excellent health.

'. . . I haven't yet begun to feel very sociable and (as
our neighbours at dinner, &c. have not yet turned up to take
meals in cabin) I have made few or no acquaintances. One
man, a Major B., who is going to Madeira, where he has
lived three years, says climate is during winter like an
English fine April day. He cultivates sugar-cane, fruits
and vegetables, with which cane he supplies Army and
Navy Stores. I got a good deal of information from him.
I have also picked up with an old colonial returning to
Grahamstown in Cape Colony (400 miles from Cape Town).
He speaks very highly of the healthfulness of the place ; he
has with him three daughters (the girls Lil noticed) and five
sons. Is rather a nice old boy. We may possibly go first
to Grahamstown and then work down to Cape: our plans
yet unformed. They all speak highly of the healthfulness
of the inland country. There are several returning to Diamond Fields who seem to find life very pleasant. There are
oceans of children on board, several Dutchmen, eight
doctors, the German Transit of Venus observer, &c. I
have enjoyed Trollope's " Africa," which have finished.

'I have nearly finished George's "Progress and Poverty."
Tell Mr. Vacher I have really rarely enjoyed a book more.
I don't agree with *all* his conclusions ; but do in the main.
His style is singularly clear, persuasive, and rich in illustra-
tion. I want you and Lil to *get it at once* (it is only $4\frac{1}{2}d$.)
and read it aloud. I have also begun three novels.

'The ship is altogether well appointed and indeed all that
could be desired. We shall certainly go on, landing at Cape
Town or Port Elizabeth according to circumstances. You
may be sure we will only do what will be the most prudent.
I feel to-day what I have not felt for months, that existence
is pleasant. We only stop two or three hours in Madeira,

and may likely not go on shore. Temperature is now 76°
in cabin ; the lowest it fell to last night was 66°. Saturday
night it fell to 52°, which is the lowest it has touched.

' Atlantic, South of Teneriffe : October 19, 1882.

' Dearest Mother,—As there is nothing to write about,
the best way to write it will be to add something daily or
thereabouts.

' Firstly I am still more all right than when I wrote off
Madeira. We came up to the Island as the sun was
setting. It looked, in deep shadow, wonderfully wild and
picturesque ; the mountains black and gloomy, but banded
with white fleecy clouds, standing against a gorgeous
opalescent sky. It was dark before we anchored and (as it
promised to be coldish and we had only three hours to stop)
I concluded not to go on shore. The ship surrounded by a
score or two of boats with Portuguese vendors of chairs,
pots, fruit, boxes, and so on. They climbed up sides, and
negotiated with great zeal—altogether an amusing and
interesting scene. We left at 10.50. Yesterday another
superb day, thermometer about 80° in cabin ; not lower than
68° all night. They had a dance in saloon last night ; five
ladies danced. Mr. Honman was one of the men dancers. I,
looking on into skylight, was amused. I talk a good deal
to father of the three girls ; he gives a good deal of information, mostly of Colony.. Had also interesting talks with
many from Natal. All say no native works after he has
saved enough to buy a wife (who does enough for both),
unless he is ambitious and wants two or more wives.
From all I hear, I think I should like the Colony much.

' *October* 20.—Another day gone. Last evening had
long gossip with colonist, giving me history of his life.
Draper's apprentice, then buyer at seventeen in Edinburgh
house, getting 230*l.* a year ; at nineteen left for London on

doctor telling him Scotland would kill him; traveller at
300*l.* a year, then on his own account in a small way.
Then to Cape (partly for his own health), where he now
employs over 100 hands, and makes 6,000*l.* or 7,000*l.* a year
and is painfully robust. Has lot of stories of dying men
who, coming to Cape, make rapid recoveries, marry, and
settle into monsters of health. Temperature last night
sank to 70°. After being 90° is now 80° in my cabin.
We sleep of course with ports wide open. I had a
delicious bath yesterday. I have read much about Cape
and am getting reconciled to idea of settling there with
you, at least for the winters, if I can't stand English
winters. The climate is praised by everyone, and there
must be some fine scenery. As Cape and Natal are five
times bigger than Great Britain, there is room enough.
I could buy a waterfall and 5,000 acres of ground, and we
could lead quite a jolly existence. I often wonder if you
would have been ill for more than three days if you had
come. I doubt if you would have been; though there are
still three or four ladies who do not appear at meals. The
last two days the wind behind us; the ship is delightfully
steady.

' *October* 23.—Nothing to report last three days. Monotonous—eating, drinking, and sleeping, but getting health
daily. Sea smooth as a lake. Flying fish, swallows and
porpoises only things in sight; not seen a ship for five
days. Sleep in pyjamas; no sheet, open ports, and panting
at that. Have had bath twice; sea water. Make acquaintances slowly only. Numerations (for Lil's benefit). A
Natal doctor, rather pleasant; went out for health (which
much improved) seven years ago. Likes climate, only
too hot weather. Speaks highly of natives, as everyone
does who has had much experience of them. Says his
only difficulty with them is that they don't like being

N

ordered about by his wife. They think it degrading
to obey a woman. Young doctor going out on spec.,
just passed at Dublin; naïf and good-tempered. An
ex-small railway contractor, now settled in Cape, at which
he grumbles. Thinks no place so good as New Zealand;
given me much useful information on railways, &c.
The " Comet," *i.e.* Herr Matsch, German astronomer, who
lives in England, and, oddly enough, has been mixed up
with Lowthian Bell, Newall, and others I know. Knows
White, &c.; is also connection of Lil's friend Helmholtz.
I find it too fatiguing to make talk, except occasionally.

' *October* 24.—Called off at 4 P.M. yesterday to join the
Grahamstown party (who make their own tea every afternoon) for their private afternoon tea. . . . After dinner
a three hours' political discussion on war, Egypt, Cetewayo,
Colonial Government, Gladstone, Bright, English parties,
law, &c.

' Parties thereto; A. (a Manchester merchant of 50 or
60) going out to see his son who has settled in Natal; the
man from Mansfield, who is an active politician, cousin
of Firth; a very intelligent Natalian who has been in
Durban since he was eight years old (and has given me
much useful information); another young Manchester
man, the contractor, &c. This really amusing. Two boys
sit opposite us at table, one son of Manchester merchant,
going out to friends in business in Natal, has been in
Holland for six months, learning Dutch; the other a very
pretty little lad of 14, son of clergyman at Cape, has
crossed five times, general favourite.

' *Sunday*.—Had church at 10.30, after a muster conducted by Captain. Didn't go. Chapel in evening to
which I *did* go. A Dutch minister on board gave four
long extemporary prayers of usual advisatory, impertinent,
and profane character; and a fearful sermon, of not bad

composition but with no point or useful end of any kind
one-third consisted of scraps of scriptural language. Told
us tempter of our souls was always walking around, and
that he was sometimes more energetic than at others, and
that he was a roaring lion! and a torrent! and a ravening
beast! and a ghostly enemy! and that we were to put on
the whole armour of faith and raise the Lord's standard!
and do a number of other figurative and impracticable
things. It seemed to me if he had told us not to gamble,
or drink, or eat too much, or cheat our neighbours, and to
help those who are helpless, and not look down on steerage
passengers, or be inflated with a big find of diamonds, or
a rise in landed estates, &c., it would have been infinitely
more to the point. The three girls and three or four men
play and sing most evenings. I sometimes go down.
There is a good [deal] of card-playing on board, and some
"sweepstaking." I don't go in for either, of course. I
fancy a doctor at Cape does well; perhaps best at Natal.
For visiting at a distance they charge a guinea for every
three miles. Thus, if patient lives 9 miles off, fee is
three guineas, &c.

'Don't get through much reading, though I think I
do more than any three others on board. Have, so far,
only read George's two books (which are all I told you
before); Trollope's "South Africa" and two other South
African books; Besant and Rice's "The Ten Years Tenant"
(a clever collection of stories); "Hades to Olympus"
(cleverish, but stilted); a little physiology; a *very* little
"Alkali Trade," and some light trifles. I am now on
Thackeray's Sketchbooks. Very interesting, and quaintly
illustrative of the line of thought of forty years ago.
His papers on French dramas, caricatures, and novelists,
very pleasant reading. I have oceans too many things,
shan't want a third. Find more than ever, if you want

a thing done, do it yourself (unless you can get such little women as Mother and Lil to do it).

'*October* 27, *Friday.*—25th, reading mostly in afternoon. Tea with the Grahamstown party, which entailed a long gossip with one of the female children. Crossed the line in the afternoon; no ceremonies of any kind. Music in evening; one or two of the three girls played a good deal of light bright music by heart. On 26th and 27th, feeling a trifle seedyish. Honman has sent me into my cabin, and otherwise tormented me, in order to keep me from interfering with him. Been reading Waterton's "Travels," "Alkali Trade," Jeaffreson's "Book about Doctors," "Voyage of the Sunbeam," stupid novel of George Reade's. Am about all right again. Talking to another Natal doctor; he also praises climate; been out eight years. A man of small capital could, I fancy, live happily enough. Eight per cent. on mortgages; nine per cent. on house property. Thermometer for last three days been between 76° and 90°; cooler than when north of line.

'*Tuesday*, 31*st.*—Had cooler and rougher weather, though thermometer not under 60°. I've been keeping pretty much to saloon and my cabin, as wind feels cold. This is tiresome; but you see I am going for over-caution. Had a theatrical performance on Saturday night; went off fairly well. Crowded house.

'. . . There is considerable singing and playing. Read a good deal; have demolished *Gulliver* for the third time; Dilke's "Greater Britain," electrical book, &c. A good deal of card-playing on board, in which Honman and I don't join. Have just re-read "Times" of October 11, for *third* time. The woman whose face mother said she liked (for reasons unknown) is a German teacher going to Cape; speaks no English. Am always thinking of you both, and of time of our meeting.

'*November* 1.—Am much better; sea continues roughish, head winds; a good deal of water shipped; but the sun has been out again for the last two days. Been talking to a young Dutchman, born in Cape, who has just returned from six months in Europe, and to a Dutch Cape minister, on Cape Dutch, &c. Also found a Middlesbro' man who has been five years in Natal; says he wouldn't go back to England on any account. Events *nil*; not seen a ship since leaving Madeira. Off Cape Verde saw some butter-flies forty miles from shore; also some swallows and an albatross, and a few flying fish. Among the second cabin passengers is a Kaffir, who has been paying a six months' visit to Europe (Rome, &c.), from money he has saved.

' . . . H. reports conversation :—*B.* "Does anyone know what Mr. Thomas is?" *C.* "A missionary, I believe." *D.* "Missionary be d—d! I reckon his mission is to make money."

' *Friday evening, November* 4.—Mail is collected early to-morrow morning, Saturday, as we arrive in evening at Capetown; so add last words. Honman recommends strongly our going from here to Calcutta. Have just had my talk to Captain; says he doesn't think we can get a steamer to Australia at all. I shall go to Australia, if we can get a steamer to Calcutta. I will send this as soon as we decide. We do not stop at Capetown, except to land passengers and mails, but go on to Port Elizabeth, and so up country, where I shall stop till I feel quite strong and well. One stop will be probably at Grahamstown, which is a town of 9,000 inhabitants, said to be pretty and healthy. I shall very likely not go to Capetown; but in this shall be guided by circumstances. Temperature for last week has been 60°–70°; wind roughish and against us. I have not once been sea-sick, although felt uncomfortable several times. Have quite got rid of cold caught in tropics. I am

so prodigiously careful, and keep in all the evenings—
which is a trial, as it is the sociable time. Think of you
all the time. Look after each other, and (Lil) see mother
has plenty of drives. It is clear I am all right in a warm
climate. Everyone is now writing letters. I shall cable
to-morrow. I look forward to getting on shore; though I
am less tired of ship than I expected.

'*Saturday morning*, 10 A.M.—Post just closing. All
right. Table Mountain in sight.

'*Later*, 4 P.M. *Extra post.*—Feeling very bright;
every one preternaturally amiable. Had games at *Words*
last night; much chat. We shall most likely not get in till
dark; we shall probably lie outside Cape Town for a few
hours. Out of seventy-five passengers, know about forty-
five or fifty. The men quite bright. Refreshing to see
land again. Passed our first steamer this morning. I
am dying for news of you all; write often. Once more
dearest love, yours,

'S. G. T.'

'Grahamstown : November 9, 1882.

' Dearest Children,—Am writing at 10 A.M. in verandah
in front of the swell hotel of South Africa. Temperature
about 60° in shade; air clear and bright and invigorating.
I well and bright also. Now to resume, from the point of
posting my letter off Capetown and cabling you as agreed.
We got to anchor in Capetown Bay, 300 yards from shore,
at about seven on Saturday, and put off our Capetown
passengers in a boat, not allowing anyone else to land or
anyone to come off.[1] Capetown lying at foot of semi-
circular precipice of Table Mountain, some thousand feet
higher, looks very picturesque; is best at night brightened
by electric lights along one quay. At 8 A.M. next morning

---

[1] This was on account of small-pox in Capetown.

we steamed away along a precipitous, fine, bold but in-
hospitable-looking coast, and had two beautiful days' steam
with wind behind us and big roller waves, keeping land
in sight; arriving in Port Elizabeth at 7 P.M. Monday.
Anchored about 1,000 yards from shore, there being no
means of coming nearer. Slept on board, and went ashore
in a tug next morning, Tuesday, at 10. Air of the bright-
est; cool, almost cold wind; fleckless blue sky, and
brilliant sunshine over all. On landing, a crowd of negro
and Hindoo and mixed porters (all colours), among whom
was one in a yellow shirt, blue vest, red turban, and
whitish pants, whom we secured; and he carried our
innumerable traps to Custom House. (Horror of horrors,
twelve packages; two-thirds at least absolutely superfluous.)
We had to open all up at Customs; then deposited all but
a bag of H.'s at the station which is on the quay. Oh, so
ridiculously English a station! A bookstall, with "Fortnightly," "Contemporary," and "Nineteenth Century;"
porters in regular English porter's uniform; carriages,
engines, cloak-room, ticket-office, &c., all conspiring to
make one think oneself in England, but for a plaintive
group of coloured folk who crowded the third-class car.
After clearing ourselves of our traps, I felt a free man
again, and recorded a solemn vow, never, oh! never, to
let anyone fix up four packages for me, or to impede me
with six others. We then located at the best hotel; a very
good one, excellent. Then called on my bank manager's
son, a bright young fellow, clerk in a big store here, a
huge place where they have stocks of Manchester and
woollen goods, wine, spirits, beer, implements, wire, tools,
and everything else. They set up country stores; have
goods on credit to enormous extent; 40,000l. worth to
one customer, they tell me. Asked younger G. to
dine with us then went to fine public reading-room

with all new books, periodicals, papers, &c.; called
at lot of shipping offices, and saw the town; two long
streets of shops and stores on the shore level, and the
residences on the hills above. G. dined with us, and
I picked up a certain amount of information from him.
He has been about two years out, likes place and
climate; costs 50 per cent more to live than in London;
profits large, but risks considerable; great drawbacks from
want of harbour works; goods landed in tiny boats, ships
often wrecked while lying at anchorage. At table d'hôte
about fifty, many not living in house.

'Next morning introduced to G.'s business man; clear-
headed. Proposed for him to act for us in a new trade.
Started by train for Grahamstown (120 miles inland);
travelled with two men from Capetown and a young
barrister, all going to Assizes at Grahamstown. Had
much pleasant chat on colonial law and customs, and
prospects and land, &c. All speak most highly of
Grahamstown as pretty, healthy, comfortable, &c. It
is a very English town. Journey took six and a half
hours. Land covered chiefly with low scrub, very hilly;
rounded rocky hills with bottoms and "kloofs" in ravine
valleys, dwellings very far apart, then little cottages with
iron roofs, or native mud houses. The third class full
of natives, and station crowded with do. Cacti, aloes, and
scrub with willows, with water, are predominant. Passed
lots of ostriches; patches of cultivated land in the valleys;
but few sheep and cattle find a home in the scrub.
Weather as before; do. to-day. This, as before, I find
suits me exactly.

'An excellent hotel; thirty rooms. About forty sat
down to dinner, I next to my steamer fellow-passengers,
Hon. P. and his doctor. We had some pleasant talk, &c
A highly-educated, well-travelled man, with rank preju-

dices; so we naturally disagreed, but pleasantly enough,
on every topic. Pleasant neighbour from Diamond Fields;
great believer in their future. In evening talked to a swell
who had been in N. and S. America, Canada, Australia,
N. Zealand, &c. Read dozen Cape papers and old
" Illustrated London News" of 1st October. To bed
at 9.30; read in bed " This Son of Vulcan," for half an
hour.

'Slept till 6 A.M., when girl brings you cup of
coffee. Snoozed till 8. Breakfast—discussion with P.,
did not interfere with consumption of four eggs and
porridge. My room fair size, high, comfortable, on ground
floor, opens out of another room. Hotel crammed. In
run from Capetown had good deal of talk with Natal
accountant and the Natal doctor. It seems one can get
7 to 7½ per cent. interest there on mortgage. Mercantile
profits very large. Not a good place for working-man
emigrant, but excellent for smaller large capitalist. The
doctor reiterating his praise of its climate. Had some
talk to the P. man; he has seen much; is ex-M.P. for ——
shire; said to have been an active Conservative member.
His science very confident and very weak. My chief
acquisition, however, a woman six feet high, whom I had
carefully avoided (by reason of) her stature and appearance.
Her son of eighteen, a thorough colonial, with her. I
found her on trial a very intelligent business woman; gave
me much information on diamond fields. She is a widow;
her husband and she had a store at Kimberley, and bought
a mine cheap; she herself used to sort the washed stuff
and fish out the diamonds. Told me much [that was]
interesting as to occurrence &c. of the stones and mode of
working. On her husband's death sold to a Company,
keeping two-thirds of shares. Net return fell from 6,000l.
a year to almost nothing. She had meantime retired to

England, bought an estate at Epping, &c. Now returning
to try and set the Co. in order, and she will, I think, do
it; wanted me much to go to Kimberley, to see her mine.

' At hotel at Grahamstown met a George St. engineer who
has been for three years at Kimberley and Natal putting
up water-works. All Kimberley people very confident of
permanence and future of diamond fields. 5,000,000l.
worth of diamonds now said to be found a year. It is
the great feature of South Africa at present. I think I
foresee other diamond fields will close Kimberley, where
they now have to go 300 feet and more below ground.
There are over 30,000 people at Kimberley still. The
nearest railway 300 miles; intervening country almost a
desert. At Kimberley nothing grows. Coal said to be
14l. a ton. Ostrich-farming and sheep divide with
diamonds the thoughts of the Colony. Ostrich feathers
worth 10l. to 30l. a pound! Ostriches fluctuate in value
between 20l. and 50l.

'*November* 10, 9 A.M.—Just had breakfast; been up
mines. Another brilliant day. Yesterday, in morning,
strolled over town. Streets immensely broad and long,
planted with trees; many good stores. Two bishops, two
churches; chapels, &c., in plenty. Magnificent public garden,
in which oaks, cactus trees, ferns, aloes, pine, firs, gum trees,
willows, roses in full bloom, pinks, and all kinds of unknown
flowers, shrubs and trees in strange juxtapositions, laid
out in a *kloof*; rocky hills above; a stream (now dry)
running through it. Streets full of bullock waggons,
each with sixteen bullocks; men on horseback or in
two-wheeled carts, with two, four, or six horses. Kaffirs
everywhere, doing the hand-work and driving, doing all
work, in fact, except that of hotel waiters. Kaffir men dress
anyhow; women in cotton gowns and bright handkerchiefs
chiefly; seem very quiet and obliging, and try to be jolly,

under not very elevating circumstances. Climate, if this is [a] fair specimen, is certainly beautiful; it was 82° in sun, 70° in shade to-day, a good deal of air in the shape of breezes. I talked to an old Englishman in charge of a pumping engine at gardens; he has been twenty-five years in S. Africa : before that four years [in] Australia ; laments over Australia, says he gets 5s. 6d. a day here, and that house rent costs him 10s. a week; everything but coarse food costs, he says, two or three times as much as in England; he says, truly I think, that Africa is no place for labourers, as native competition too [severe] here; but great place for capitalists. He had been twice to Diamond Fields, but did no good either time. I walked about three miles yesterday without fatigue ; no pain ; cough only two or three times in evening, if I get in cold air.

' Am just starting for a trip to lower river for two days, in style ; have joined another in hiring a trap and *four* for our two selves there and back, so shall have easy time. I still think we shall have to get to Calcutta in order to make our way to Australia.'

' Grahamstown : November 13 (Monday).

' Dearest Children,—I resume at point where I left oft my last, viz., as I was starting in the two-wheeled cart and four, specially chartered for the occasion. Myself, H. (an " Africander," or descendant of Dutch settlers, and secretary of a Capetown bank), and F. (a Capetown civil servant magistrate, born in Colony). Driving out of Grahamstown by a fine road, all up-hill, had pretty view of town, with its many trees and churches ; " City of the saints and city of woods," covering much ground in a depression, with hills all round it. Over brow of hill a great expanse of hills and valleys, sea in distance ; hills and valleys alike parched-looking, though a few clumps of trees in valleys

At twelve pulled up and took out horses at roadside
hotel, standing almost four miles from nearest white
dwelling. Met many natives on road, and saw many of
their huts, also abundance of cattle and some ostriches
(showing more in the grass than one would think).

'Starting again, up-hill and down, bumping and jolt-
ing; country began to be greener, grassier, and a lot of
bush and small woods, chiefly of a tree looking like a thirty-
foot aloe ; only heard native name, blister tree ; it is only
an overgrown plant, and doesn't look like a tree hardly.
After passing through a village we got into a pretty little
gorge, and debouched on the River Couri here, about half
a mile from the sea ; a fine tidal river, with wooded hills
running down to the water, making it not unlike Dart-
mouth. I think the port is destined to be an important
one when harbour is finished. At present about one hundred houses, mostly galvanised iron, on the hills on the
two sides. We crossed by big ferry boat, and drove up
to hotel on brow of hill ; going up, the cart stood on its
back, chiefly.

'The hotel outside looked like three galvanised iron tool-
houses, all in the last stages of decay, and stood together to
prop each other up. However, on going in, it improved
vastly, and we arranged for lodging : I am getting a sofa-bed
in a comfortable little sitting-room, and the other two a
room ; river between them. Before dinner, strolled down
to harbour works, which consist of pushing out two pieces
of concrete blocks of fifteen tons, between which river runs
out ; they have not yet got to the right point, and vessels
have to lie in open bay outside the bar, as in all S.
African ports but Capetown. We watched tug going out,
seas breaking over her from stem to stern. Dinner at
6.30 ; to my surprise five men to dinner besides ourselves.
Bank manager, harbour master, a swell settler, and two

others. Dinner excellently cooked, and good, though simple. Sat and talked in verandah till nine, when went to bed.

'Next day very hot; lounged in morning, in afternoon got boat and two men to row us up river for five miles. Pretty; the wooded rounded hills coming down to water. Some of the houses of the town looking over river very pretty, but all galvanised iron roofs and generally ditto walls, with· rough brush inside; when painted white looks all right, otherwise it only looks—good for trade.

'Sunday morning we started back here at 4 A.M., just sunrise, which was lovely. Came by a different and prettier road, country covered with copses of aloes and different unknown shrub-like trees. At eight, stopped an hour for breakfast at a nice little inn in a sort of tiny valley, with a pool and spring in it; a garden full of bright blossomed flowers, and a first-class breakfast of eggs, coffee, and minced meat. . . . Then up a fearful hill miles long, and with sun beating down like a fire, and back to Grahamstown at 12.30 ; distance each way about thirty miles. I am clear that the way in summer here is to get up at daybreak, sleep from eleven to four, and work in evening again.

'To-day I have moved to a very nice large airy room across the way, mealing still at hotel. It has been raining, more or less, all day; rain greatly wanted, and is the more appreciated now, as it comes in a soaking sort of drizzle, and not in a tropical downpour. Had long discussion with P.

'With my comrades to the C's. I had much interesting talk. F. a very intelligent, well-read man. We discussed Comtism, natural theology, Darwinism, the native question, the Dutch influence in the Colonies, their civil service

and magisterial system, &c. &c. It appears that there
is a resident magistrate and civil commissioner in each
district, who is a civil service clerk really; his only training in law being what he can pick up as clerk to some
other magistrate; they have most extensive powers; can
sentence to a year's imprisonment and fifty lashes &c.,
and are receivers of all crown revenues. F. has lived in
various districts, on which he gave me much information.
He says that Kaffirs meet with much injustice, and are
often very badly treated; they are also subject to a number
of very onerous regulations; cannot move without a permit, can only rent land; they are in fact made a modified
kind of serfs. A young farmer who has been here two
years was fiercely arguing with me that they ought to be
allowed to shoot natives whenever they saw them tres-
passing; he finally wound up by saying, "Well, when we
do it, we are always acquitted by a jury;" which is unfortunately true. I listened outside a Kaffir church on
Sunday; very earnest singing and preaching, in most
emphatic, eloquent style; sounds much like Welsh preaching. The preacher a native.

'*November* 15, 10 A.M.—Yesterday and the day before
were wet and drizzling, cold; the thermo. not under 60°;
however, kept in doors chiefly, only going up to the
library, which is a fine public [one] and very well supplied with books, periodicals, and papers. I have been
reading up my "Contemps," "Fortnightlys," and "Nineteenth Century." Have been talking much to P.; we
disagreed mostly about all things; but he is intelligent,
well-read and travelled. I have induced him to read
George's book, which horrifies him beyond measure.
There are about seventy sitting down to meals daily. I
attended two sales yesterday. One of a farm (including
house of eight rooms) of 3,000 acres, which sold for

2,100l.; it was within twenty miles of a good port, and less of a railway; suitable for birds, sheep, and cattle. I should have bought it if it had sold for 1,700l. The other, a little seven (rather insignificant) roomed house and garden in Grahamstown, let for 50l. a year, sold for only 320l., repaying about 12 per cent. Have been talking to N. (member of Legislative Council here) on native commission. Attended Court of Session. The native is here put on a theoretical equality, but practically far from it; thus, in Court all the seats occupied by whites, natives standing; so if native assaults white, heavy sentence, if white assaults black, trifling one or acquitted. Natives all dress here; some, particularly women, very well, but nearly all bare legs and feet; women generally a bright coloured handkerchief on their heads. To-day lovely after the rain; cloudless blue sky; bright warm sun. They tell me that further inland at Cradock it is often 140° in the sun and 105° in shade; but that it does not feel oppressive even there.

'I talk to every one I can get hold of, and read all the numerous local papers diligently, and am coming to know a good deal of local conditions.

'*November* 15, 4 P.M.—Wet again; yesterday fine and hot. I spent morning walking about town, attending Courts. Heard native tried for cattle stealing, and very properly acquitted. Then to library, back to lunch at 1. Then to post-office, short walk, and long read at library. To-day, walk before breakfast; then called about town, inquiring about investments, &c. Lunch, talking scandal. Among my new acquaintances, a storekeeper at Cradock. Gives me much information on up-country life—Cradock being the present termination of railroad, and Dutch. Dutch are the great conservative and obstructive elements, oppose all improvements, whether railroads, water-works,

bridges [or] fencing, &c. Storekeepers' profits universally
admitted to be very high;' yet here I see in windows
trousers 10s. a pair, girls' waterproofs 7s. 6d., &c. On
the other hand, little things 50 per cent. to 100 per cent.
dearer than in London. Talked much, too, with L., a
London mechanical engineer who came over here for his
health six weeks ago; has settled his wife and children at
Cradock, which he praises much for healthiness; it is very
dry, treeless and dusty, 3,500 feet high. Gives wonderful
accounts of chest invalids who have recovered marvellously.
He thinks of settling here if he can get any engineering
work. Have just been talking to a man, a born colonist,
who has very large farms 40 miles inland; said to be most
successful farmer in Colony. I have taken great fancy to
him. He speaks well of Kaffirs, if you look after them;
pays them 11s. a month and daily allowance of 2 lbs.
meat and 1 lb. mealies; they save money and sometimes
own up to 30 or 40 oxen, which he lets them graze on his
land. He has cattle, sheep, and birds; says birds pay
best, but require much care. Is fencing all his land; says
it is indispensable, as ostriches otherwise will run away,
40 miles in a day. Ostriches give 30 or 40 chicks a year.
Says English farmers coming here lose money from doing
everything in English way. P. and I had much talk
again; his little Jersey doctor also very confidential. It
appears P. was specially recommended to come here by his
London doctors.

'November 17, 8 P.M.—Honman turned up from Capetown last night, very pleased to have seen his sister; had 500
miles to go each way. He says he has heard of several prac-
tices vacant, and to be obtained without payments, which
are worth over 1,000l. a year. I think Arthur might, on
passing, do well here; everyone says that a sober doctor
does exceedingly well. It would, however, be necessary to

learn Dutch to do really well. Have had no word or line from you of any kind yet, since I left; begin to want news badly, please always keep press copies of your letters; I may have two chances of getting them. We spent this morning in gardens with C. girls; walked this afternoon about five miles in all.

'My breath, cough, and chest, all *very much better indeed.* I still think it would be no hardship to live here, if it were not for patents and researches. I am clear I should be all right here; but I am not sure if this or sea suits me best. We can hear nowhere any tidings of direct steamer to Australia. We shall probably go *viâ* Natal. I quite think life here could be tolerated very easily. Am really *much* better. Thinking constantly of you. Wish I had mother's photo. in my triptych; send it me. Take care of each other!—Ever yours.'

'I may probably not be able to get off another letter for at least two or three weeks. Please keep my letters, they may serve as signposts hereafter.'

'Grahamstown: November 22.

'Dearest Children,—Still no news of you, which bothers me much; otherwise all right, but decidedly tired of the monotony of this.

'We unfortunately have no introductions here, so have no one but the C.'s. Spent Sunday afternoon there. Their garden is divided from the public Botanic Gardens by a stream which is now perfectly dry; after rains is 6 feet. In the garden are orange and lemon and fig trees in full bloom; with pears, plums, peaches, strawberries, cherries, pumpkins, also loaded with fruit though not yet ripe. Roses, fuchsias, and geraniums, with aloes and cactuses, abound. Their house a long low one, only one story

O

high, with big cellars underneath and verandahs, furnished in newest English style.

'Our life very monotonous. I am up about 7.15, breakfast at 8, then to post office always to find *no* letters. Then to library, a stroll, luncheon. Then sit in balcony, library, stroll; dinner at 6.30, general chat, and to bed about 9.30 or 10. P. and I have long talks on all kinds of subjects. He has a lot of introductions here, so gets asked out a good deal by the resident magistrates and the merchants here. Another character is a man named W., of an Anglesea county family, who says he knows everyone, and has been in Canada, U.S., Australia, and New Zealand. He is now going, and for a year's shooting expedition. Shooting, &c., up country with a bullock waggon.

'Then we have a German from Diamond Fields; has been there twelve years; made and lost a fortune; full of regrets for Germany, dislikes the country much. The Diamond Fields are a worse locality to live in. If a man buys a diamond from a native or from anyone not a claim holder, he is liable to a fine of 5,000*l.*, twenty years' imprisonment, and a hundred lashes. This is monstrous; but is constantly acted on.

'We went for a short ride to-day; threatened rain, so we soon came back. It rains more or less every third day. Everyone here complains of bad trade, absence of money, &c. At the Diamond Fields things are certainly much depreciated; shares in the diamond companies having in every case sunk to one-third or even one-tenth of their value a year ago. This depression at Kimberley reflects itself even here. Thus the carriage of goods to Kimberley is an enormous industry; goods are carried from here in bullock waggons, carrying four tons, drawn by sixteen oxen, at a rate of from 36*l.* a ton in good times to 16*l.* a ton now.

' As fresh men come to hotel daily from different parts
of Colony, I collect and compare views and facts from
varying grounds. There are several large schools here,
particularly a big Church of England Grammar School and
a big Wesleyan Girls' School. The natives have no good
school here, but have one some forty miles away. It is
said that the Kaffir is particularly bright at mathematics,
and when initiated in Euclid and Algebra, spends his play
time in working original problems.

' *November* 23.—Another mail in ; still no letters. I
am getting desperate, and cannot even be consoled by my
six " Times " up to October 26, which I gloat over at the
Library.

' *November* 24.—Your letters of 25th and others just
arrived—such a relief. I had been wiring about all over
the place to get news of those letters. I have also news-
papers &c.

' It was wet yesterday, so we did not go for a ride,
for which I was thankful, being " stiff " to the verge of
distraction.

' To-day we have been with C. to see camera obscura
of some friends. People interesting, and views of country
and town wonderfully perfect and curious.

' It is possible we start in a few days for either
Calcutta or Australia, but the sailing is so uncertain, we
may be kept some time. P. talks of going up country in
a bullock waggon. I should much like to go to Diamond
Fields, but give up as they say they are unhealthy. You can
hardly imagine how I enjoyed your letters, and how much
I look to the meeting.

' Am all right, but still rather " scant of breath ; " think
sea will make me a finished cure.'

'S.S. " Moor," off Port Alfred:
'November 30, 1882,

'Dearest Children both,—Am on the move again, so
feeling happy. Sunday, Saturday, and Monday last nothing
happened, but *weather* showery and uncertain. Made a
few fresh acquaintances; had several long talks with P.,
and one long ride which I enjoyed; had fine canter on the
downs above the town, which we and horses enjoyed alike.
I found, however, holding on rather wearied me; we then
rode all over the native location.

'The natives live *entirely* out of the town in about 700
huts, each with a small piece of ground which they
cultivate. In most cases they have bought the freehold;
in others pay the Government 1*l.* a year rent. The huts
made chiefly of wattle and mud, but some of galvanised
iron. In the immediate neighbourhood of their houses
the aborigines dispense with a good deal of superfluous
clothing, in which they have my entire sympathy; they
also think that a good many of our so-called necessaries of
civilisation are really superfluities. By the way, their
regular wear is a garment which they dye of a highly
æsthetic dull brickdust colour, which suits alike their
complexion and surroundings. On Sunday, however,
they go to their kirks in the most elaborate English
costume. They have one chapel in the town, another in
their location.

'After a long consultation with H. we decided: (1) That,
though Grahamstown was a good enough place in its way,
it, and in fact all the South African health resorts, were
too high up to suit me. I find that, though I have quite
got rid of pain in my chest, which was the main and really
dangerous business, my lung is only improved very little,
being worse than when at sea. (2) That, as we can't get

to Australia, we ought to go to India in the cool weather,
stop there ten days, then on to Sydney. Having passed
these resolutions, I began to feel better at once.

'Tuesday we paid farewells to C. P. had also discovered that Grahamstown didn't suit him much, so determined also to leave this week. He is going for a long
waggon trip up country with his doctor and servant, and
is to let me know his experiences. He is actually looking
far worse than I, who indeed present a robustious appearance. He has had fracture of the skull, broken leg and
ribs, and several other trifles, but fully expects to go back
to Parliamentary life. Must have had a fine constitution;
tells me for years he never took more than six hours' sleep.

' Wednesday morning we started by rail for Port Elizabeth. Miss C. and H. and his sister came to see us off.
We travelled down with a young Scotchman we met
at C., named Hamilton. Pleasant fellow; much talk;
has been [here] over five years. He and a brother, having
5,000l. each, bought a wholesale saddlery business at
Port Elizabeth and Grahamstown for 10,000l., and have
been at it since. Profits about 40 per cent., or say 100
per cent. per annum gross ; but great risks, as they have
to give four months' credit to people in remote parts of
the country. As an illustration of risks, a bank here,
which has just smashed, has lost 20,000l. in the Transvaal. Everyone says the Dutch here are utterly opposed
to all progress. In the Transvaal our retirement has
been followed by a sort of general bankruptcy, and they
are at their old occupation of pillaging all the surrounding natives. At Port Elizabeth living is very dear. A
clerk can hardly live on less than 150l., while salaries are
relatively low,—130l. to 250l. I have talked to so many
men from all parts, that I feel I know South Africa in-
timately. When we got to Port Elizabeth at 6.30 P.M.,

I found that there is a fine French ship, the "Havre,"
sailing for India from East London; do not know if she
is taking passengers. Found also a Union steamer leaving
for East London at once; so charter a boat for a sovereign,
and arrive on board the Union ship "Moor," a far finer
boat than the "Conway Castle." We are now lying off
East London, and I feel as jolly and bright as can be. A
sea life suits me, I think, and hill air does not.

'We hope to get to East London this evening, and either
to get a passage in the "Havre," sailing about to-morrow,
or in the "Clan Cameron," sailing next week. I only
propose stopping in India about ten days. Read L.'s
article on George in the November "Contemporary." I saw
the magazine in Grahamstown more read than I ever do in
London. Money is going fast; at Grahamstown we paid
25s. a day for our joint boarding. I stayed on at G. till
Wednesday, hoping to get a letter from you by mail leaving
London on November 2. Out! though I got one from
Per.,[2] I did not from you. You can have no idea how I
appreciate seeing your writing as a sort of physical liga-
ment with yourselves, of whom I am constantly thinking.'

'East London : December 2, 1882.

'Dearest,—I closed my last on the S. S. "Moor," which
brought us here at 7 P.M. on the 30th. We were landed
in a tug in complete darkness at 8, and found our way up
the hill on to the plateau on which this town, the third
seaport of South Africa, is built. We had some difficulty
in getting put up; though there are forty hotels here.
Finally got into what is said to be the best, but it is a
woeful falling off from Grahamstown; rooms dirty and
cooking indifferent. Yesterday interviewed the agents and
captain of French ship Havre, but he won't take us, as he

[2] Mr. Gilchrist.

carries no saloon passengers, and we can't go steerage. Honman called on the leading doctor here, who was very polite, drove him round town, &c. I had a short chat with the doctor too. He came here six years ago, and H. thinks he must be making over 1,200*l.* a year. The doctor says there are a number of places in South Africa where a good steady man can live and make 1,000*l.* a year, this being one. H. is quite bitten with desire to return here and do so. The doctor had bad health in England, lungs weak, and had to spend winters in Madeira, so came here.

'There are four doctors here. Population of town itself 3,000 or 2,500 whites; it has a railway, and it is a considerable port. I don't much care about the place, it is so intensely new; three-quarters of the houses all galvanised iron, dusty, hot, and windy. Talking to three or four young men, who all gave dismal account of colony; had all been to Diamond Fields, and all more or less failed; two going home again. Also long talk with a colonist born here who has large wool-working establishment up-country. He, as every other English colonist, complains much of bad feeling and jealousy of Dutch population, who oppose all progress and improvement. He too has been at Diamond Fields, says land here is too dear; in his part, which is a feeble part, it fetches 30*s.* an acre; thinks it may be further depreciated, &c.; says natives work well for living, their only fault cattle-stealing. A wife costs ten oxen, and these they think it their duty to raise from somewhere or somebody. All the rough labouring work here done by natives, artisan work by whites chiefly, and the looking on business done by whites solely. Had ramble over downs this morning. Afternoon (heavy thunder showers) spent in public library. Been interviewing shipowners. A plague of flies here, and a few sanguinary and persistent mosquitoes; shall be very glad to be on board ship again. If we can

get a ship for Australia, shall take it, but can't hear of one
so far. As instances of prices, we paid 25s. a day at
Grahamstown hotel for the two ; here we pay 20s. a day ; of
course all drinks extra. Thus, a pint bottle of zoedone,
which costs 6d. in London (or 1s.) costs here 1s 9d., a pint
of English or German beer costs 1s., a pint of champagne
17s. 6d., and so on. Here, especially, the bar is always
crowded with young men, clerks, business men, &c. having
brandies and sodas (1s. 6d.), gin slings, sherry and bitters,
&c., on which they must consume a quarter of their income.
All say that most of the doctors drink.

' *Sunday, December* 4.—Went to bed at 8.30 last night.
Have a three-bedded room ; but this fellow (a Mr. Bell) is
rather an acquisition, as he tells us a great deal. Thinks we
treat natives both most unjustly and stupidly, which seems
quite the case. We are always stealing their land and
pushing on boundaries, but do not govern in any proper
sense. He says they make good workmen if well and
fairly treated, and that they prefer being struck for a fault
to the intervention of magistrate. To-day it is blowing
quite a gale, so I am stopping indoors, as H. says. Wind
cold, though sun so bright; had a smart thunder shower
last evening. Have got a sitting room to ourselves. There
is every likelihood of this becoming a big place ; but it
has first to go through some vicissitudes. So wish I had
a photo. of the Mother; the one of the saucy-looking child
is quite a resource. It is possible that we may still go to
Australia instead of India.

' *Thursday, December* 6. *Noon.*—Still at East London;
have just decided that we will go to Mauritius and Bombay,
thence to Calcutta. There is no chance of getting direct
to Australia, and H. thinks a long voyage the best thing.
I believe I shall have always to remain near the sea. I
don't suppose we shall be more than ten days in India.

We shall probably get to Calcutta between January 12
and 20. H. says I am much better than at Grahamstown,
though I was well enough there. The only trouble is that I
cannot walk more than a mile at the time. Find it very
dull here, though we are at the best hotel (which is less
clean and more flyingfied than it might be) and are members of the Club. H. has made the acquaintance of a man
named Pyper, cousin of Dr. Cotman. He is a clerk in a
merchant's house there.

‘Had long talk yesterday with a man who has spent
seven years in the interior of the Zambesi. Says it is
fine country but unhealthy; that a pass from Matebele, who
is chief of a district 500 miles broad, secures from any
hostility of natives. Buffaloes, elephants, rhinoceros, and
ostriches very abundant. Gold, lead, &c. found but not
worked; as waggons only practicable part of the way, must
be on foot.

‘I look forward to our four days in Mauritius. Besant
and Rice's novels make one feel to know it. I find it very
trying not to be able to make walking excursions, but
caution is the order of the day. We are going for a ride
this afternoon. Expect to sail from here Saturday, but
may be delayed, and (if so) shall not get to Calcutta till
near end of January.

‘If Arthur passes next summer I might perhaps bring
him out in winter and settle him, if I have to come away
myself, which I hope I shall not have to. We get short
cables daily from England, generally about three lines.
Thus, yesterday had news of Arabi's trial, death of Archbishop, relapse of Trollope. This climate would suit the
mother gloriously; sunshine from morning till night, with
generally a cool wind, sometimes a cold one. H. dined
last night with P.; before refused to. He (P.) lives with
three other men; they rent a five-roomed house for 6l,

a month ; have two servants, one black girl gets 12*l*. a year, the other English (cook) about 30*l*. Rent is everywhere enormously high, as are luxuries ; meat 6*d*. a pound, butter 2*s*. to 3*s*., eggs 2*s*. or 1*s*. 6*d*. a dozen. Just going to post, thence to Club.

' Yours ever affectionately.'

'East London : Friday, December 8.

' Dearest Ones,—Have just, while wasting in despair, received yours of November 9. Don't now send any but technical papers, please. The photos are indeed a treasure ; I would rather them than considerable pelf. Mother not good, but still good enough to be a treasure. . . . I wrote yesterday, saying that we sail in " Clan Cameron " for Mauritius, and then probably to Bombay, and on by easy rail stages to Calcutta. I have arranged, however, that if we get to Mauritius in time for steamer to Australia, we have option of joining it. There is, however, very little chance of this. I am longing to be at sea again. This is excessively dull ; the high winds, heat and dust, prevent our riding or walking, and we can't leave on account of the uncertainty of the ship sailing. Nothing happens but picking up a new acquaintance, going to Club or Library. Had the editor of local paper with us last evening. Had amusing talk with an old Italian ex-captain, as agent of an Insurance Company. His verdict on South Africa : " This is no fine country at all. This have much dust, much wind, no water, no food fit to eat, no nothing at all." It does not seem on the whole much appreciated by the residents.

' I fear there will be a " war " against the Basutos shortly. It is really a pillaging expedition, the farmers openly saying the object is to confiscate all the land and cattle.

'The great trouble here is that anchorage is so bad
and exposed that ships are sometimes two months un-
loading. On Sunday and Monday, all the steamers had
to leave their anchors and steam out to sea, so losing three
days.

'*Saturday, December 9.*—We sail this evening. Had
young fellow to whom H. had introduction to dinner; is
in stores, been here three years, says even a clerk does
better here than at home. He came out on spec. After
three months' waiting, got a berth at 12*l*. 10*s*. a month,
now 20*l*. a month. Says clerks and principals rise much
more here than at home. His housekeeping with three
others costs him 8*l*. a month. Says no society here, no
dances. We went yesterday to a place few miles away;
pretty, but absence of big trees painful. The Euphorbia
is practically the only tree here, and mimosa bush the
prevailing shrub. The winter here has cold, often frosty
nights, but bright sunny days, with frequent cold winds,
but frost in day time not known. I regret not seeing
Natal, but it would entail going on *viâ* Zanzibar, which
is unhealthy.'

This is the last South African letter, and we may
interrupt the correspondence for a moment to observe that
no one would imagine from reading these epistles how
seriously ill Thomas really was. He writes indeed with
the vivid energy of a man in full health. The contem-
poraneous correspondence of Mr. Honman with Mrs.
Thomas gives, so to speak, the reverse of the medal, and
brings into relief the dark background of deadly disease
which lay behind the superficial gladness of these travel
days. Mr. Honman writes from Madeira of bad nights,
pain in the sides, and frequent coughing—'heavy fits of
coughing.' At Madeira, however, the cough is 'of a

better character,' and 'the worried look has to some extent disappeared.' 'I hope,' says Mr. Honman, 'that with care at the Cape, he may be able to take some pleasure when he gets to Sydney.' On November 3, there is further good news: 'Sidney, in spite of dull cold weather, is better both in health and spirits; he has, I noted, attempted to part his hair to-day—not a very successful effort, but a most favourable sign; he is particular about his collars as well.' There has been, however, more 'pain in the side,' and an 'attack of pleurisy.' He is as careless as ever of 'himself, and will talk to anybody in the coldest wind.'

Improvements continue to be spoken of in Mr. Honman's letters from the Cape; but the warning note is still constantly recurring to one reading between the lines. There is 'great shortness of breath,' much greater at Grahamstown, however, than at East London.

The result of the letters is, substantially, that although symptoms change and soften in character, the lung trouble never really disappears.

In Thomas's own letters there is naturally a constant desire to make the best of things for the sake of the anxious ones at home, whom (as appears from every line he wrote) he loved so dearly. He made indeed ineffectual attempts to 'edit' the communications which he knew that Mr. Honman was making to his mother. The good doctor writes from East London on December 8: 'I have no doubt he has given you all news, but I write this and send it separately, as he desires to revise my letters.'

CHAPTER XVI

MAURITIUS AND INDIA

WITH this necessary interruption we resume Thomas's correspondence with his 'children.' The reader will be able to sufficiently discount the praiseworthy affectation of good health which he will occasionally detect.

To his Mother and Sister

'December 14 or thereabouts, Thursday.
'S.S. "Clan Cameron," Indian Ocean, Lat. 295°.

'Dearest Ones,—I posted you a letter on Saturday morning last at E. London, and one posted on Thursday, and went on board at 3 P.M. on Saturday. At E. London, you must know, no ship can come nearer than half a mile from the shore; so we got out in a little tug which tosses and tumbles considerable, and we, (H. and I,) clamber on board by a rope. There is no shelter, and the stormy winds do blow with praiseworthy persistence and force. So the "Cameron" has been ten days putting her cargo on shore; this being done by lighters. We soon found that we should not start that afternoon, and it began to blow in the evening and continued all Sunday; so that the lighters for the balance of cargo could not come out, and we were pitching and tossing at anchor in a most distressing way. Monday morning, however, was decently calm; so we landed balance of cargo and got off at about 2 P.M.

'The ship, a new iron one of 2,400 tons, on her second

voyage only. Captain pleasant, chatty little man, who
has hitherto commanded Australian passenger sailing
ships chiefly; only his second voyage in steamer. Only ac-
commodation for eight saloon passengers. Comfortable
saloon and cabins; only all too near the screw, which is
a peculiarly noisy one; fare and attendance very decent.
No doctor on board, so Honman is sort of semi-official
honorary surgeon. A doctor came out in the "Cameron" to
set up in E. London, but after ten days concluded to give
it up, and returned to England the day we left. I was
sorely tempted to leave H. behind at E. London, as he
would have liked; but I feared you would raise some
paltry objection and get alarmed if I did, so I heroically
brought him along. Am I not quite too good? . . . I,
too, eat like a hale and hearty crocodile.

'Now for our co-passengers; to gratify Lil's morbid
curiosity. No. 1, Scotchman brought up as working
engineer; in '70 working in Manchester at 36s. a week;
found his master would only screw more work out of him
the more he did, so determined to try Kimberley. Started
within a week of hearing that good work to be got there.
Started there at 4l. 10s. a week; lived on 15s. a week.
Helped a man who wanted to import machinery; taken
on at 9l. a week. Presently started on his own account
as small engineer; got a partner; worked up business,
turning over 80,000l. a year; then amalgamated with a
larger firm doing still better. Is now taking trip to
Australia, America, England, and back to Kimberley.
We talk of his experiences, the Fields, engineering, &c.
He is really nice fellow to know. I have taken quite a
fancy to him; he does not boast or swagger, but is full
of information. Has just been showing me his collection
of photos and stones &c., including nine rough diamonds.

'No. 2 is also from Kimberley, a Scotchman, making the

tour with No. 1; has a store; has been in America thirteen
years, gold-digging, in N. Zealand, &c. &c.; very pleasant
and intelligent. No. 3 is on his way to Australia, thence
by United States home. Affects the swell; has been
twelve times in U. S., also in China, Japan, India, &c.
No. 3 forms with No. 4 a hostile camp. No. 4, young
engineer, has been five months in Kimberley. He and
No. 3 (who has only been six weeks in the Cape) abuse
Colony all dinner-time 'every day till No. 1 can stand it no
longer, and mildly points out that all their facts are wrong
and their conclusions without foundation. I naturally
support No. 1.

'No. 5, young Swiss, been eighteen months in Cape
trying to open up business in Swiss goods, but has not
succeeded; takes it philosophically; is now going to Ré-
union and then to join a firm in Madagascar. His father
has factory in Baden, Switzerland. He has spent a year
at Birmingham and is fairly bright generally. No. 6 is
a young fellow from Glasgow, who is going the round
trip for his health. Started from Liverpool, and goes viâ
Mauritius, Bombay, and Suez Canal, home by same ship.
I believe they charge 90l. for the trip, which will take
about 110 days to 120 days. Besides these, there are four
coolies who have been a year in Africa and are now going
to Mauritius, where they expect to do better; and a family
of German Jews who are abandoning Africa as not suitable
for tailoring enterprise.

'We sedulously do nothing all day long. I have read
"Celia's Arbour" and "Monks of Thelema" since I came
on board; both very amusing. Besant and Rice certainly
have more in them than ordinary novelists; they always
work in some queer social ideas and are unconventional.
Have you read "George" yet? Mind, I shall examine
you both in him severely. Did I tell you I made poor P.

read it? which was rough on a proximate peer and Irish
landlord. His criticisms, however, were fun, and clever,
and kept me on the *qui vive* as defender of the faith of
George. The day I left Grahamstown P. accidentally
told me his views of me by saying that my views seemed
to be nearer those of *Cowen* than anyone else. Soon after
he described Cowen as a man of the most odious and dangerous views, though &c.! I miss my talks with him; it
was exciting sparring sometimes, and kept one alive. . . .

'*Monday, December* 18.—Have had beautiful weather
ever since my last, hottish, but never over 85° in the shade,
and generally a cool breeze; am certainly benefiting
much. H. says I am getting quite fat-faced. We all get
on well together, talk and read. I read chiefly, but talk
considerably with M., who has shrewd ideas on subjects he
knows. . . . I have been reading of California; it seems
after all *the* finest place in the world for climate, fertility,
and everything together. Taking it all round, I think
there are a number of better places than S. Africa. I feel
now that I know *all* about S. Africa, and could pass an
exam. in its resources, politics, sociology, climate, &c. We
all long to get to Mauritius, to have a run on shore. I find
we cannot get to Calcutta before January 26, or thereabouts, which means about March 15 for Sydney, and not
leaving Australia till end of April. In short, I hardly see
how to get home much before July; but all this may alter.
I should like a fortnight in America, if I come that way.
It is a dreadful time to wait before seeing you. The
photos get constant attention.

'*Mauritius, December* 22.—We anchored off Port Louis
on Tuesday afternoon, having been for two hours skirting
the island, which has several ridges of most romantic and
striking looking precipitous mountains, some running
straight up from the sea, some springing from the interior

lowlands. Port Louis has a superb situation, being backed by an amphitheatre of hill and precipitous cliff, with the slopes covered with low thickets, vividly green, with great patches of scarlet flowers. We anchored a mile outside the harbour, and (to our horror) were put in quarantine, with a talk of being kept for two weeks. At 5 P.M. on Wednesday we had the joyful news that we were allowed out of quarantine, but concluded it to be too late to land, so landed on Thursday after breakfast. The situation grew more striking as we neared the landing. There were some thirty vessels in the harbour, a busy quay, and the town white, but embowered in trees. Our ship, from the moment quarantine was removed, was invaded by a multitude of boats, all manned by Indian coolies of multifarious races, and Chinamen or Malays; numbers came on board —such handsome men; some of the Malays and Indians in the loveliest linen garments, and scarlet girdles and turbans, forming an extraordinary contrast to our ragged dirty crew.

'On landing we spent two hours or more walking in the town. The market, a very large building, crammed with Hindoo and Chinese vendors, with a few negroes and half-castes, but not a single white person. Shops mostly kept by Chinese or coolies; many of stalls kept by women in most picturesque costumes. The effect of a bright green under-garment covered by an overskirt or burnous or thingumbob of scarlet, and a few brass (or gold) ornaments, is delicious. Lil should adopt it; it might be necessary for her to improve her personal colouring with walnut juice. Also, I know no more becoming dress than a white linen nightgown with scarlet sash, deep collar and cuffs. This last I propose adopting myself for Chelsea and office wear. Blue is the only colour they seem never to use, except the Chinese. It was my first experi-

P

ence of Oriental life. There are certainly fifteen different shades of colour and race; the rarest in the town being the pure negro.

' As we shall stop here till 26th or 27th, probably till 29th, we have come up in a body to Cureppe, the sanatorium of the island, 1,500 feet above the sea level. We came by afternoon train, making a party of seven—we being now excellent friends all. We break up here, three to Australia and United States, one to Bourbon, one stops here for a time, and one goes on with us. The railway ride was very interesting, picturesque, and strange. The coolies, who thronged stations, peculiarly interesting. There are 250,000 coolies here, they say. The women work in fields, carrying loads, &c., to a painful degree. They come here on five years' contract, but generally stay on. For further description of Mauritius, its scenery and customs, see Besant's and Rice's Xmas number of I. L. N., " Ready Money Mortiboy," and " My Little Girl."

' On arriving at Cureppe, we, after a reconnaissance, descended—the seven of us—on an hotel kept by French people (everyone in the island nearly is French; nearly everyone speaks French, though mostly English as well). The hotel is in a large garden, running over with palms, tree ferns, aloes, roses, bougainvilles, pine-apple, shrubs, and 963 other flowers and trees, for which, if I invented suitable names, the mail bags wouldn't hold the list thereof. It is such a contrast, too, to Cape hotels in living and bedrooms. Here everything scrupulously clean; there all scrupulously dirty ; there, bad cooking and doubtful food ; here, French cooking and delicious fruit, salads, &c., with lovely coffee. Breakfast is at ten, and dinner at seven. The village is all round exclusively coolie and Chinese shops, or as nearly so as possible the shops about ten feet square, some only five feet.

' It has been very hot to-day, so much so that I found
a walk of 500 yards quite enough; but in the morning
and evening it is quite cool. All the servants here are
Indian, deliciously quiet, swift, and efficient. A Hindoo
watchman keeps all night in the verandah just outside
my window. The watchman, with his turban, toga, and
bare legs and feet and staff, is a highly picturesque feature,
though I fail to see his utility.

' *December* 24, *Sunday.* I am wearying of the intense
idleness of the life; yet, it is impossible to do anything.
The moisture of the air makes one feel an insurmountable
languor; though temperature only 85° in shade. There are
constant tropical showers, and it *does* come down when it
comes. I have been several short walks round. The ground
fertile to a degree, and crowded with the quaintest and
most variegated types and colours of people. *All* the
shops are kept by Chinese. . . . We went this morning to see people coming from church—oh, such smartness
and colour! The nights here are regularly cool, though
mosquitoes a little troublesome. . . . It is all French here;
only one waiter understands any English, though all
servants are Hindoos; our host a thorough Frenchman;
hostess and daughters ditto. The number of travelling
hawkers (chiefly of cakes, sweets, &c.) is surprising.

' We have not succeeded in getting any English papers
here, so I know nothing of English affairs since November
8. I look forward to getting to Calcutta with the utmost
anxiety. I must say I am very weary of idle wandering.
We can't get any saddle horses here, which is a great
disappointment.

' *December* 25, *Xmas morning.* Thinking much of you.
Up at seven. Very bright and sunny. Thermo. down to 70°
and people coming from church in gorgeousest of raiment.

The bulk of our party going an excursion, from which
I have cried off.

'Yesterday fetched a pleasant ride through the Botanical
Gardens, &c. There are charming houses, chiefly verandah
and garden all round, this being the residential quarter.

'*December* 28. Xmas Day, went to church at nine.
Church dressed with palm branches, ferns, and flowers;
crowded with white folk, well dressed; thirty or forty carriages waiting outside. A number of coloured folk of all
hues, in back seats and standing; the coloured girls all
have white muslin thrown over their heads, looking very
picturesque and well. Sermon in French, singing not
first-class. Mass of usual elaboration, gorgeously dressed
attendant boys.

'Breakfast at ten; lounged in verandah till one, when
H. and I, and one or two Kimberley friends, started to
drive to the waterfalls, three miles off; passed on our way
hundreds of the coolies' huts (wretched hovels of boards
and thatch mostly), and thousands of their occupants,
children, and brown and black in all shades, in all degrees
of non-clothing, but mostly plump and well formed.
Passing a sugar mill, we stopped and went all over it,
finding it very interesting. The canes brought to the mills
by two wire tramways and a traction engine, in addition
to endless trains of mule-carts, bringing the cane to the
rollers direct from the fields in which it was cut. The
mill is, of course, tremendously hot, as boiling and eva-
porating is going on all over the place. All the work is
done by Indian coolies, who work very hard. Their average
earnings are only 4*s.* a week, and rations worth 2*s.* a
week more. At the factory gate was the usual Chinese
general shop, where we bought some Scotch beer and some
soda water and biscuits, by signs chiefly. They kept
everything you can think of, but their great business is

in dried fish, rice and rum, which the coolies and natives
buy in ha'porths and penn'orths.

'The cascades are a fall of a small river some 400 feet
in seven falls, very beautiful in their way. They fall into
a deep, profusely wooded gorge ; precipitous peaks tower
on either side, and then the gorge opens out by a further
fall into a rich plain of sugar plantation, bounded by the
sea.

'Yesterday (27th) we all reposed. On 24th had been to
crater of extinct volcano, a mile or so from here; very
curious and romantic. All the soil here is lava, the
whole island being of volcanic origin. . . .

'*January* 1, 1883.—On 29th bade adieu to our friends
at Cureppe, and came down by midday train to Port Louis.
Spent two hours in luxuriating over a fortnight's "Times,"
bringing us up to Nov. 20, and got on board our old friend,
the "Clan Cameron," after spending half an hour in
hunting all over the town for photos of the isle, which we
failed to get. We found the captain and mate ailing from
Mauritius fever; they having been on board in the harbour,
which is hot and unhealthy. The young Scotchman from
Glasgow had also stopped on board, and was also ill; but
none very bad. Found, to our disgust, no other saloon
passengers to Bombay, so we and the young Scot are all
by ourselves. We had pictured the pleasure of having
engaging young Mauritiennes as co-passengers. When
we got on board, the ship crowded with some two hundred
coolies, of every shade and type of face, saying their adieus
to forty coolies and Chinamen, who are going with us as
deck passengers to Bombay, having served their five years in
Mauritius; the Chinese are *en route* for Hongkong. There
are two women and a baby also with them. The ship
is rather heavily laden with sugar for Bombay, and rolls
heavily, taking in water all the time, which makes it

wretched for the sick folk, and less comfortable for us; they are, however very good-natured over it. Several speak English, but more French (or rather a sort of semi-French-English-Hindostanee).

'*January* 3.—We have had two lovely days and this is yet another. The thermo. ranges between 75° and 85°; sea calm, and motion of ship creates a pleasant cool breeze; so that on *deck* it is never too hot, but just the enjoyable temperature, though when we go to bed it is too hot to sleep till 1 A.M. or so. . . The officers are not strong in conversation. We all now sit at one table; but I and the Captain have to do.all the talking that goes on. It seems that, on ordinary sailing vessels now, chief mates only get 8*l.* a month, second 7*l.* or less, and third 5*l.* 10*s.* to 6*l.* You can't expect a man to talk much on such a salary as these. Of course on steamers the rates are higher.

' Respecting the general dearth of conversation at table, I stumbled last night on a great joke. I said something about it being hard work to keep some talk going to the Scot, when he replied, " Well, you know, I think you are some restraint on them, Mr. Thomas; I don't know if you prefer·not being addressed by another title!" I puzzled my head for explanation, which arrived at, amounted to the fact that all the officers and men, having maturely deliberated, have concluded that I am Sir Gilchrist Thomas, Bart., and have been observing my movements with great interest and curiosity in consequence. W. had written home with a description of the affable Bart. The myth seems to have originated in that source of all evils, Lily's dreadful calligraphy, her "Sid." Gilchrist T. being read as " Sir." Please Lil take this as a warning. . . .

' Have been reading Haweis' "Current Coin" (which you should get); they are suggestive rather than thorough [essays] but bold and advanced enough for a clergyman. I

have been having some square thinking on religious questions, partially led thereto by Lynton's "Under which
Lord?" (which also read); it is clever and trenchant, and
à propos to the times, if occasionally overdrawn a little. . . .
    'I collect testimony when I can as to efficiency of Hindoo
labour. The general evidence is that two good Indians
are more than equal to one good white man in most kinds
of work; while the wages of the two are less than the
wages of the one by at least one half. I am full of fresh
ideas and experiments I want to work up and try. I am
inclining to leave business alone as much as possible.
    'January 5.—We crossed line yesterday evening in the
loveliest of weather; the 84° of heat being tempered by a
slight breeze increased to a pleasant one by the ship's
movement.
    'This morning is close, damp, and oppressive. . . . I
am picking up a deal of nautico-commercial and ship-
building information. The Captain has been thirty years
at sea and twenty-four in command, always sailing ships
till last voyage. . . . He sticks to his opinion that New
South Wales is the best place in the world. I am
wondering if the mother could stand a trip to California
if I find it wise to go away next winter; H. thinks not.
When sailing about as now, always remaining (when on
land) in English ground, one feels pretty strong symptoms
of pernicious British pride. I read and think in a desul-
tory way a good bit, and don't feel very bad at the con-
finement. If I had you two with me I should be quite
content. I am now anxious to get to Calcutta for news.
    'Sat up late last night reading a book of Thomas Cooper
on Christian Evidences. Have been examining Bible and
Prayer Book to-day with great diligence. The skipper
came up and looked over my pile of books and, to his
great astonishment, found:—1. Prayer Book; 2. "Alkali

Trade ; " 3. Cooper's Book; 4. Electricity; 5. Bible; 6.
" Cleveland Engineers ; " 7. "Iron;" 8. Novel; 9. Blue
Book on Australia. He wanted to know if I read all those
at once; to which I of course replied that I did.

'The sunrise and the sunsets are glorious; after all
cloudland is a picture gallery open to all which it is not easy
to surpass for loveliness of form, colour, and every changing
variety. It is, however, always dark by seven. One misses
the twilight. I have got into the way of waking for the
sunrise and then going to sleep again. . . .

'January 9, '83.—Off Bombay. We hope to get into
dock in about an hour; we are however rather late for
the tide, and may not get in this tide. Will reserve my
impressions of Bombay till they are consolidated. Our
run from Mauritius has been a very quick and pleasant
one. With the exception of one muggy, windy day, the
weather has been glorious; the thermo. never fluctuating more than four degrees on each side of 81° day and
night. During the last few days nothing of any kind has
happened, beyond once sighting a vessel six miles away,
which is but a mild form of excitement. Conversation
has languished, though we are all on the best of terms;
there is simply nothing to talk about. . . . I am feeling
well and bright; no pain for a month, cough a little in
evenings occasionally. Can read and think well. I am
going to stop away so long to make betterness permanent.
Eat prodigiously. I long for letters. I haven't spoken
to feminine human being for six weeks. . . .

'Bombay, January 10.—Landed at 5 P.M. yesterday;
the ship lying in the harbour. You see very little of the
size of the town from the harbour, which is spacious and
sheltered. We missed the tide, and so couldn't get into
dock. We drove through native town for some three
miles to the Adelphi Hotel. The town crammed with

humanity ; yet somehow does not convey idea of a city
with 770,000 inhabitants. The natives live largely in mud
and thatch open hovels, giving no protection, it would
seem, against rain. Hotel a huge place, two-storied, with
enormous verandahs and galleries. . . . Sat in verandah
reading old papers and being bitten by mosquitoes. . . .
We go on towards Calcutta this evening, stopping two
nights on way, Allahabad and Benares. I shall not
bother about Agra and Delhi (much as I should like to
see them) as they are out of our line. We shall be riding
about Bombay to-day and getting money, &c.

'This morning we have had successively visiting our
bedroom (which has no glazed windows, only wooden bars)
coffee-boy, newspaper man, barber, boot-cleaner, bath-man,
washer-man, and a few others. Crows and pigeons abound.

'6 P.M. Just leaving. Had a pleasant day.'

'Benares, 7 P.M.

'*January* 13, 1883.—Dearest Children,—My last
left me at Bombay on Wednesday, when, after calling at
Bank, H. went down harbour to see a friend on another
steamer. As he did not return for three hours, I chartered
a boat and five, no one of whom understood a syllable of
English ; and at last glided triumphantly down the harbour to the " Clan Cameron." I said good-bye to officers ;
found H. had been there, and got back again in triumph ;
chartered a cab, which here is a first-class vehicle, something like a Cape cart, or a high hansom cab with the
driver in front, and drove all over town, chancing on H.
driving in another cab. One of my searches was for a
sun helmet ; but Bombay could not raise one big enough
by three sizes.

' We started for Calcutta at 6.30 P.M., being seen off by
our only co-passenger on the " Cameron " and having only

one other in our carriage, which was a first saloon to hold nine, and sleeping arrangements, .with water, &c. laid on. We had a very comfortable night, though we required our rugs. Our fellow-traveller [was] a native, who had bedding, crockery, glasses, milk, fruit, dressing cases, and every conceivable appliance. He made himself very friendly and obliging, spoke English well, and gave us much information. In the morning a grand sunrise, still cold; country all day varying between a fair state of cultivation and monotonous scrub or semi-wooded ground. Village of mud and straw huts, miserable looking to a degree, scattered about at rarish intervals. We passed through some fairly pretty nooks and valleys in crossing the Ghauts, which are quite low.

‘ At 8 on Friday morning we got to Allahabad and got into excellent quarters at the best hotel. . . . .We . . . hired a carriage and a guide for the day, and went to Public Works Office, where I saw several polite officials on business, and got some information ; then through European and native towns, which are quite separate, the former consisting of tree-planted roads 100 feet broad, with stucco semi-classical buildings standing back in grounds (shops and private houses alike); the native town, narrow streets lined by little open shops, no fronts at all, each five to ten feet square ; in most cases manufacturing and selling going on in the same contracted space as carpenters, smiths, potters, brass-founders, image-makers, jewellers, cap-makers, sweetmeat and cake makers, fan-makers, fiddle-makers, and a score of other trades. Vegetable, fruit and stuffs sellers are almost confined to the bazaar or market. Then to see the great bridge carrying the railway over the Jumna, just before its junction with Ganges ;. this junction is a specially sacred spot for bathing in, and for the next few weeks millions come to bathe. From the bridge (a brilliantly

designed iron structure) to the fort (which commands the
actual junction and a fine view); it is largely garrisoned,
but we got a pass and went all over it and the stores and
workshops—these last entirely operated by natives.

' *January* 14.—In the fort is also a long subterranean
cavern or passage, with irregular niches, occupied by
images of gods, which were being worshipped by peram-
bulating crowds. Atmosphere abominable from crowd; no
ventilation, and grease lamps.

' At dinner thirteen at table, one a lady in white evening
costume. Not having seen a good-looking woman for three
months, I couldn't keep my eyes off her. Talked to my
neighbour (a male, alas!) about native servants, who cost
four to ten rupees a month; horses, including two grooms,
cost four rupees a week. Then got talking across table to
General Napier Campbell, a man of about fifty-five. We
had a long talk about literature, politics, America, &c.,
continued after dinner in his room; very pleasant and
intelligent, as evidenced by his saying he had enjoyed his
conversation.

' Went off at 8 A.M. this morning by train to Benares.
We had to cross river on bridge and drive four miles to
the hotel, which is, as usual, in the European quarter;
then drove back to the town in tow of a regular guide (for
the first time in my life am I so degraded). " Fergusson " [1]
took us, however, regularly round to about ten superior
temples, mostly poor enough architecturally but quaint
and barbarous to a degree; some laid over with gold
plates, but mostly stone or plaster covered with red paint.
The Monkey temple, colonised by some hundreds of
monkeys of a sacred herd, who seem fully as intelligent
as their cultivators, struck us as perhaps most curious,

[1] It is probably unnecessary to say that the allusion is to Mark Twain's guide ' Fergusson in *The Innocents Abroad*.

and I was intensely interested in the monkeys. In the
Golden temple, which is crowded with sacred cattle, and
has a well (which smells like a bad sewer) in which the
god Siva resides, we were mildly mobbed on the question
of offerings and backsheesh. We were, after four hours'
templing, tramped through the bazaar streets, which I
really enjoyed more. The crowded way, *jammed* vitality,
and yet impassive unchangeableness of the life is at once
interesting and oppressive. I hardly suppose the native
towns, or way of life, or arts, are changed from their state
two thousand years ago.

'A pleasant party at dinner, though no general conversation. Four ladies—an event!

'Up at six next morning and drove down to the river,
where we met "Fergusson" with a boat, and we rowed up
and down for two hours, watching the thousands of religious bathers. The whole side of the stream is lined
with stone steps or terraces, some fifty or a hundred feet
high, surrounded by magnificent buildings, built by
different rajahs to commemorate their visits to Benares.
The steps and terraces themselves covered with minor
shrines, idols, &c., and thronged with multitudes of the
devout going down to, or coming from the water, or
standing in it. Men, women, and children, in blue, white,
red, green, mauve, gold, yellow, violet, crimson, purple,
and every combination of all or any of these human in-
genuity could devise. I could have rowed up and down
all day, but "Fergusson" insisted on depositing us at the
Railway Station an hour before time, and (after selling us
some fraudulent Brum. coins at ten times their value, and
charging us preposterously for his services, leaving me
with one and a half rupees in my pocket and our tickets),
he insisted on our giving him a gaudy testimonial. . . .

'In the next compartment was some almighty swell in a

green nightgown, blue pants, and gold vest, pink and gold turban, silver and gold shoes, turned up and coiled over in rings, in case his feet grew during our journey. *Item* : two infants of ten or twenty ; one male, one female ; very fat, in green, gold, white, red, blue, and silver satin. *Item* : two soldiers. *Item* : two silver sticks, and one gold ditto in waiting. *Item* : six coolies to carry their bags. *Item* : three superior and six inferior officials ; chief duty to give lollipops to junior swells. These infants must be even more spoiled than our silver-spoon youngsters. We rode on, having a fine compartment to ourselves, dining and supping *en route* gorgeously, and sleeping comfortably wrapped up in rugs at night. When at Benares and Allahabad it was quite cold. Thus, on Tuesday, it was 135° in the sun at 2 P.M. ; at 2 A.M. it was only 50° or 43° in the grass.

'We arrived at Howrah, the "Surrey side" of Calcutta, at six A.M., and drove over here, which is the swell hotel in the swell street. . . . On entry, found a whole host of servants waiting to be engaged. Finding it is usual here (as we had been warned) to employ one or two servants each, we took on our table servant, and a majestic man in silk and white linen, with a white turban, began to take off H.'s boots and hand him his hair-brush, which greatly gratified H., and we presently found that this great being had engaged himself as body servant. They then chevied the balance away, leaving us quite alone, bar a gentleman who wanted to wash [us] and our clothes, a second who had shaved me before I had considered the question of being shaved, a third who insisted on measuring me for a pair of trousers, a fourth who wanted to sell me a hat, a fifth who left a silk dress on approval, a sixth who in tones of tender emotion wanted to cure me of corns, a seventh who began cutting my hair, an eighth who wanted

to take off my boots, and the six men who were getting a
bath, making the bed, and dusting the chairs.

' After breakfast we went to P. O. and Bank, and got
your three letters, "Truth," "Graphic," "Ironmonger,"
and " D. News." So delighted to get all, and above all
to hear you are both well, which is the great news. I
should say that our. guardian in the turban feels it his
duty never to leave us. While at the Bank and the
P. O., we tried to dodge him by a side entrance, but he
had us in custody again in a second. At a shop I again
nearly got clear away, but was captured after a few
minutes of freedom. This afternoon, by great fortune, I
found a cab with no seat or step for our custodian, and we
at last succeeded in escaping by keeping at a gallop.
While enjoying our freedom, saw Patent Agent, Patent
Secretary, Public Library, &c. On our return we felt
awfully penitent as our Mentor took us in charge, and
reproachfully brushed Honman down (I declined to be done
anything to), and fixed up our chins,

' Our table servant is arrayed in gorgeous linen vestments,
with a girdle and white turban. My first three evenings I
always felt dinner, with one of these silent mysterious beings
behind every chair, to be a solemn and oriental ceremony,
and I always expected to hear one whisper, either that
Fatima, captivated by the piercing glances of my eagle eyes
would a word with me in the sheltered alcove, or that
" Haroun al Raschid deemed it best for the benefit of my
health and my chances of Paradise to encircle my lily neck
with a steel wire bow-string." However, now that I have
two mysterious children of the Caliph or some other Eastern
gentleman among my personal retainers, I am getting to
feel bold and commonplace. I also feel none the worse
for journey, and generally jolly. Lil seems to have done
everything possible in business way.

'*Bengal, January* 22, '83.—Last Tuesday (15th) I called on B. and H., two members of Supreme Council— very well received ; lunched with H., his wife, and a R. C. priest or bishop ; all pleasant. Next day called by appointment on H. again, to meet six of the government engineers and heads of Public Works Dpts. I think, after two hours, I pretty well converted them all. It seems, however, that the Indian Secretary at home, has, since I left, been attacked in the House and by English ironmasters, so that there is likely to be much difficulty and probably failure, owing to interference of the English control. Tuesday also got your letters ; much enjoyed them ; so glad to hear you are both well. . . . Was asked to dine with H., but as Honman didn't want me out in evenings, refused.

'Wednesday and Thursday called on head of Geol. Survey. Very kindly received. . . . He and everyone says drink is the curse of Europeans, and real cause of two-thirds of illness. Called on engineer of leading railroad ; had pleasant chat. . . . I think the interview will bring business to N. E. S. Co. Then saw agent for Rothschild, and (later) head here of Great E. I. Rail. I found in all cases most pleasant reception ; got lot of information, enjoyed talking to rational men again. [Saw] also Secretary of Bengal Government, a Major. Of course I didn't walk a step. I had a two-horse cab, my own footman, and the cab footman hanging on behind, and all this luxury for about 1*s.* or 1*s.* 3*d.* an hour. Got your first batch of papers, "Graphic" and "Truth," but none by Tuesday's mail ; enjoyed them very much. . . .

'On Friday came down Barrakur, where the B. Ironworks are. S. (who is a German engineer looking after Works for government) met us with carriage at station, and drove us up to his house ; very nice one on hill ; climate here delicious, coldish even, at nights, about 70° in shade during daytime.

Country round not very pretty but fairishly well cultivated. S. and his wife very hospitable and pleasant; we have driven about and seen a good deal. It is a little too hot for much walking from 11 to 2, but very pleasant in house even then. Am always bright and fresh. S. four years in India; married two years ago. Was two years in Scindia's employ as engineer in general to everything. He tells many curious stories, and I have heard much of the ins and outs of Indian society. They have twelve servants here, costing 16s. a month each for wages. All keep themselves, and all are men; so the total cost is about 110l. a year. Two gardeners, one coachman, one groom, one undergroom, one top man housemaid, one under male housemaid, one water-carrier, one man for cleaning, one tailor, and one miscellaneous man. Food cheaper here than at home. A fowl costs 3d., beef 1s. a pound; a cow costs 5l. for best kind. S. and I know many Germans mutually.

'We leave here to-day; see H. to-morrow; leave Calcutta about 29th. We stay perforce ten days in Ceylon, then on to Sydney. I have enjoyed this country jaunt, after hotels, very much. First night jackals and wolves singing all round kept me awake, now I am used to it; jackals cry like the spirits of departed teething babies. Have seen no snakes, though said to be abundant. I look in my boots and hat every morning, but to my great disappointment have failed to find one.

'*Tuesday, January* 23, '83.—Here we are back again at Calcutta (Great Eastern Hotel); we had a drive, &c. yesterday and came down here with S. by afternoon train. I shall see the Council to-morrow or to-day, and then live a very quiet week here, not going out in evening till 29th. The temp. here very equable and pleasant; by no means too hot. S. says India is excellent for chest complaints, and that he has quite got rid of one chest disease he suffered from. Travelled down with a large party of English, including a young female who smoked cigarettes in a disgusting manner. I think young women who smoke cigarettes should be burnt alive, with tobacco as fuel. We took one of our servants with us to Barrakur, but finding him no use sent him back. I gave him 2s. a day (his proper wage being 1s.) and he has now turned up again with his former colleague in new clothes of the most gorgeous description from head to foot (at least he has no clothes on his feet), including blue turbans and scarves and lovely lace petticoats ; they are now both quite too beautiful to expect to do anything ; but as they never did anything before but put on H.'s boots and hold his comb and toothbrush till he wanted it, it don't much matter.

'I learnt much on Indian manners and customs at Barrakur, and am very glad I went there ; it was the pleasantest trip we have had. I really feel my mind and knowledge of peoples expanding so rapidly that I am obliged to let out all my hats. Some of my things have gone back in a box to H.'s people, with a lot of his, though he still persists in carrying his top hat along in a special hat-box, about which I keep him worried by constantly starting up and saying, "Now I believe that hat-box was put in the Simla train as being certain to belong to the Viceroy," or suggesting that it has fallen over or got sat on.

'*Great Eastern Hotel, Friday, January 26, '83.*—Dearest Ones,— . . . Saw H. for two hours, Tuesday, and some engineers and bankers Wednesday ; always in, easier to work, and meeting the pleasantest receptions. . . . Yesterday went over Geological Museum and spent some time in their library very pleasantly, the head of the Survey being my cicerone. I drove there, and am ashamed to say came home in a palanquin or palki, carried by four men. It is

Q

a curious sensation. They are much used here, and are
absurdly cheap. You can hire one for five hours for 1s. 8d.,
English money. I gave my men ten annas (the fare being
three), and they immediately started a hubbub of the first
magnitude, thinking I must be insane. H. and S. are
much exercised by my ruining servants, coolies and porters
&c., by what they call my reckless extravagance. I tipped
about ten servants at S.'s house. I began with 1½ rupee
(say 3s.) each, but got down to 1s. at the end. It is
curious that being liberal don't seem to be appreciated.
Thus, I gave coolies who put our luggage (only three bags)
in train 1s., and they asked for more. S. (who had three
big boxes) gave 2d., and was overwhelmed with bowing.
. . . Don't let mother worry. I am getting along
beautifully. I feel more and more that I would not have
missed this initiation into Asiatic life. By the way,
Keshub Chunder Sen gave a lecture last Sunday to an
enormous mixed audience, on Christianity, Natural Religion,
Brahminism, and the relations of Europe and Asia. Read
it if you can get hold of it. I was very sorry to be unable
to hear it. Did I tell you the Baboos (or writer and
merchant class) look exactly like Romans ? Many have
classical firm features, hair cut square over foreheads, and
wear a toga, and no head covering or trousers of any
kind. Julius Cæsar, or his facsimile, cashed a cheque for
me the other day.

'Both our old servants have returned to us, as well
they might, as we pay them over double the usual rate,
that is, 2s. a day each instead of 1s. The bearer is a fine
tall high-caste Hindoo ; so high that all the waiting he
can do connected with food is bringing us our early
morning coffee at six. He then gets our bath, folds up
our clothes (to my great annoyance), and gives H. each
article of clothing, makes beds, and supervises us with a

critical eye. For the rest of the day he does nothing, unless I invent an errand for him (which, as I am in constant communication with government, I often do).

' Wages at Barrakur for labourers are about 3d. per day. For women (who work, they say, often better than men), 2d.; for children (who I am sorry to say work hard from eight upwards), about 1d. On other hand, an English foreman who would at home get at most 12l. a month, gets there 30l. I went to two collieries. At one a native manager very courteous, intelligent, and obliging; gave us all figures asked ; [at] one an Englishman, also pleasant. Miners earn about 9d. a day; bring up about two tons a day each ; much less than our own men. They won't use gunpowder, owing to some prejudice. I should extremely like to push iron-making in India ; even if it cost me money, it would be a grand thing to keep ten million rupees annually in India. . . .

' I go to dinner with H. to-morrow; sail Monday. Went over Mint yesterday. Temperature beautiful ; about 65° at night, 70° to 72° in day, shade. The cruelty to unfortunate oxen used as beasts of burden is dreadful, and is the only thing that cools my ardour to relieve India of her burdens. Have seen boat to Melbourne I thought of going by, but don't like it; so shall go by P. and O. to Ceylon, and so to Sydney by "Paramatta." Have just been seeing two Ministers again ; they are frank enough, and if Kimberley doesn't put spoke in wheel, shall do well. Now for five weeks of absolute quiet, and monotony, and dulness.'

This is the last letter actually written from Indian soil (although the next epistle describes a farewell dinner), and it will be well to give Mr. Honman's view of his patient's health during those journeyings and negotiations with the

Indian Government. He writes to Mrs. Thomas from Calcutta :—

'There is a decided improvement this week in his lungs. . . . Those pains that have been so much cause of anxiety have not been present for the last month.' But the anxious physician goes on to complain of Sidney's broken sleep upon mail nights, and to urge the importance of keeping from him the details of business as much as possible. ' Will you see that everything that can possibly be kept back (unless of vital importance) be kept back ? He dreams of fresh complications each time, and he awakes with a bad headache.'

In point of fact the success of Thomas's discovery and the commercial undertakings which had followed in the train of that success had brought the usual penalties with them of much care and trouble.

A week later Mr. Honman writes (still from Calcutta):—

' Sidney has told you about the trip to Barrakur, I suppose. It has a beautiful climate at this time of year, but it is too cold at night to continue there. .... Sidney can work a great deal better than he could before, but I endeavour to prevent him as much as possible, as I notice it does not improve him. . . . The stay in India has not been such a bugbear as we anticipated. His lungs have improved since we arrived, and I have no doubt would have improved more if we had stopped longer, only I am afraid of the work here. The government people have no regard for anybody's health.'

CHAPTER XVII

CEYLON, AND THE VOYAGE TO AUSTRALIA

So, with a little rift showing in the gathering clouds, Thomas sailed for Australia. We resume his correspondence on shipboard.

To his Mother and Sister

'S.S. "Teheran," off Madras : February 3, 1883.—Saturday.

' Dearest Ones,—I left off in my last just leaving *en grande tenue* for dinner-party at H.'s, who is the equivalent perhaps of our President of Board of Trade (or nearer to French Minister of Public Works), and Member of the Supreme Council. There were eight men there and four ladies ; the men, a R. C. dignitary and military and civil servants. I talked chiefly to H., who told me his experiences of natives, among whom he has, he says, many intimate friends. (He speaks several Indian languages.) . . . Says they produce excellent mathematicians, engineers, and architects. He is an architectural amateur himself. We then spoke of ironworks &c. I am very desirous to aid in introducing these in India ; it would ultimately keep in India nearly a million sterling a year, which is now a fearful drain on her poverty.

' I talked also much to Colonel S., the Director-General of Railroads, who was born in India, was through the Mutiny, and knows much of the country. Thinks all but a small class of Mahomedans and ambitious spirits are

content with our rule &c. Also with a man who had
been " Resident " in Scinde and other native states, a very
able man and good talker. Enlarged my views on many
Indian topics; we had some pleasant sparring. He very
sensitive to English criticism and that of men travellers.
I was kept by H. after others. Dinner very good, not
ostentatious; six servants in picturesque costumes and
gorgeous turbans fastened with magnificent aigrettes.

'Next morning up at 6. (Honman, like a brick, had
done packing previous evening.) Started at 8; backsheesh;
got on board P. & O. s.s. "Teheran," a fine boat which
takes us to Colombo, where we wait ten days, sailing
again in " Paramatta " about 16th for Australia, where we
should arrive about March 13. We sailed at 9. Such
a crowd of friends to see us and the fifty or so passengers
off. Sailed down Hooghly; chiefly striking for tropical
vegetation and the enormous number of ships lying in tier
after tier. . . .

'Among other passengers a S. American, who speaks
French and is a sort of Commission to Australia, for some
mysterious purpose, studies *l'état social et communal, agri-
cole et industriel*, &c. Talks well, only the strain of French
breaks me down.

'Then there is a man named P., in the Chinese Con-
sular service, who is quite interesting; is one of a dozen
men who talk and write Chinese with perfect ease and
fluency. Gives one a very different idea of the Chinese
from that one derives from books. He says that actually
there is no religion at all among the male Chinese, though
they believe in a future state, in which, however, they do
not suppose their conduct here will affect their position.
He also describes them as highly logical and reasonable in
argument, &c. Says opium trade is a grievous ill that
we have forced on them; that it seriously affects health,

conduct, &c., of huge districts, and that Chinese are sincerely anxious to stop it. He has been lending me some notes of his on Chinese law and the paternal power.

'*Item* : A Scotchman who has lived twenty years in Boston and Toronto, made his fortune and tried to settle down in London, but had had to start round the world for a change ; has been doing Egypt, Syria, and India ; is going on to China. He likes Canada much better than England ; says too, Canada can absorb any number of really working immigrants. Has been recently in Manitoba, of which he speaks highly as to its futurity. . . .

'I am tired of shipboard again, and am so looking forward to getting home ; the long spell from and to Australia will be very tedious. The chief officers here get 20*l*. a month, the junior captain 400*l*. a year, the senior 1,000*l*. Doctor gets 10*l*. ; Honman says doctor also gets numerous fees. By the way, S. played the zither delightfully. I like it much better than piano ; it is low and melodious, and doesn't obtrude itself on anyone. . . .

'*Thursday, February* 8.—We arrived in Colombo last night ; shall stop up at a sanatorium near town till "Paramatta" arrives. We are two days late here, owing to some defect in engines, which we had to stop four days at Madras to cure. We lead the usual uneventful life. . . .

'I am always well enough ; the only thing I absolutely do not get clear of is a little cough. I often think if you could have stood the sea (which you couldn't) how jolly we might have been together. I am very savage at having to stop so long at Colombo ; we shall only get five or six weeks in Australia.

'Everyone says that no one ought to go to India after early March for the first time. November, December, and January are the best months to go. It seems certain that Europeans cannot colonise in India ; that is, after three generations in India, they die out. On the other hand, Lewis, Sir J. Phayre, and all authorities say that a man who eats moderately, drinks *not at all*, and protects his head from sun, is nearly as healthy as in England. Liver complaints are very little known. The climate seems most fatal to children; then to women. If brought up in India, they say only one soldier's child in nine lives to twenty-one; on the other hand, in a female orphan asylum, where they live with extreme simplicity, and great attention is paid to cleanliness and exercise, they have wonderfully good health. . . .'

'Mount Lavinia Ground Hotel, Ceylon : February 12, 1883.

'Dearest Mother,—We landed at Colombo at ten on morning of 8th. The town, with its red-tiled houses and clusters of palm trees coming down to sea, looked bright and pleasant. We drove about for an hour, walked through the markets &c., and I felt I knew Colombo. Many of the buildings are the old Portuguese houses and forts transformed.

'We finally came up here by train; the railway skirting the shore all the way, with cocoanut palms, among which the native houses are scattered thickly on the other side. Mount Lavinia is only seven miles from Colombo, but said to be much healthier. It is a knoll of rock, only some fifty feet high, jutting into the sea; the hotel an ex-governor's country house; it is very large, of classical architecture, and very commodious and magnificent. Thus, the dining-room is a magnificent hall, some 100 or 150 feet long, with two rows of pillars down the sides, with a number of little tables, exquisitely laid out with linen, glass, and flowers, making a more striking *ensemble* than any I have ever seen in any hotel anywhere. The dining-room opens by wide (always open) doorways into the drawing-room, and that on the verandah; thus we have the ocean on one hand, the palm forest on the other. We have a room which can take forty or fifty, with only an average of eight; though yesterday twenty or thirty came over to dine. The meals are appalling in their variety, frequency, and richness, and the cooking far ahead of anything I have ever suspected. We fare sumptuously if we take three out of nine courses. We have not wandered more than three miles away on either side. The Ceylonese or Cingalese are a fine, if somewhat womanly race; don't affect clothing above the waist; wear long hair and tortoiseshell combs. . . . This is quite a Castle of Indolence, even worse than the steamer. We revel in " Punch," " World," " Truth," " I. L. N.," " Graphic," " Pub. Opinion," " Field," and " Queen " (alas ! I have read all these twice through), and are in all ways in pampered luxury. The native fishing boats, six feet broad, twenty long, with an outrigger, are an endless subject of curiosity. They sail magnificently.

'*February* 14; *Mount Lavinia*.—Yesterday we spent in Colombo, wandering about, and chatting to some of our old steamer acquaintances. We are the only ones who have been out here. Colombo tremendously hot; but grass always green, which, after arid deserts of India and Africa, a great refreshment. We came up again in the evening, and now find we do not sail till midnight to-morrow. I shall post this before we sail. They say the " Paramatta " is a splendid ship. Our life here dreadfully slow; there are two young women here, but both married, and with their friends, who are not sociable. I get on here well enough—reading and lounging, and playing chess (for Honman's delectation). Mind, I do not believe an idle life is good for anyone at any time, and I loaf under protest. Our Spanish-French passenger from

Calcutta goes with us to Australia. We may very possibly
land at Melbourne and go by train to Sydney.

' I like the natives; they are quiet, dignified, and well
featured, though I fancy somewhat idle. The hard work
is done by immigrants from Malabar. Remember, you cannot possibly get another letter from me before the end of
April, when I trust I shall be on my way home. Tell me,
Lil dear, exactly how mother is always.—Yours lovingly.'

' February 25, 1883 : P. & O. s.s. "Paramatta," Latitude 24.

' Dearest Children,—We parted with some regret from
our palm-forest and marine-palace of Mount Lavinia early on
the morning of 15th ; went down to Colombo, sending our
traps on board. We parted, to amuse ourselves in our respective ways, till ship sailed in evening. I, lounging in
hotel verandah, soon picked up some of my " Teheran "
friends. Several were leaving for China the same evening,
among them my Chinese consul. With him I drove to
the museum, far away from the town, and saw some interesting carving, inscriptions, and jewels of old Ceylon.

' Talking (which we did at a great rate, my consul being
an interesting and aggressive conversationalist) we spoke of
Arabi, and I said I had a mind to leave my card, as a mark
of sympathy. P. jumped at this, and said he should
like above all things to interview the Pasha. We finally
compromised by agreeing to leave cards, and leave it to
A. P. to say if he could see us or not. This we did. Arabi
sent out to ask us in. His house is a moderately comfortable sort of European-Indian house, in a longish garden,
in the suburbs of Colombo. We found Arabi and another
pasha sitting on the verandah, with seven or eight sub-
ordinates round. We shook hands and began to exchange
complimentary remarks through the medium of two very
atrociously bad interpreters. The consul, to my disgust,

said I was a member of Egyptian Committee (which I am) and a leading pro-Egyptian and pro-Arabi politician! This being floridly translated, Arabi began bowing to me, putting his hand to his heart, and insisted on my taking an armchair by his side, and showed me an elementary Arabic-English phrase book, in which he was grinding up, pointing out words " *my friend* " as describing me, and by bowing, smiling, &c.,,conveyed his goodwill. I felt rather an impostor, but disclaimers proved no good. We continued to be cruelly mistranslated, and to be obviously made to say imbecile things, till I was reduced to the verge of distraction; but the consul was quite equal to the emergency. Finally coffee drunk with infinite empressement, and a loving parting.

' Arabi looks earnest and determined, but does not strike me as peculiarly brilliant; *not* a very striking face, but still beyond the average.

' After excursing further about Colombo, and having a final gossip in the crowded hotel verandah, I went to our ship in one of the native outriggers, which are the queerest but safest of craft.

' " Paramatta," as you will have seen in papers, is a fine new boat—this has led to her being very crowded—there being over a hundred saloon passengers.

' A young pair only two or three months married; the husband, quite youthful, is going out as first Professor of Anatomy to the New Medical School at Sydney. His wife still young, pretty. The professor is well up; speaks French, German, and Italian, and knows some general science. To my great astonishment I found, after a day or two, that Mrs. R. and Mrs. Professor between them have persuaded the man whose cabin I shared to retire in favour of Honman to a far inferior cabin, leaving his to Honman and self. . . .

'It having got abroad that I am with a doctor, and there being nothing visibly wrong, it is generally supposed I am a dangerous lunatic. . . .

'We have an ex-Victorian merchant, now living in Tasmania, of the healthfulness of which he gives the most glowing account; an ex-Sydney merchant; a N.S.W. surveyor, born in colony, returning from tour round world all by himself; an ex-ship captain who has recently lost his wife, taking voyage to get over his loss; a missionary who sits next me at meals; in intervals of eating (he consumes more than I should have thought physically possible for anyone), answers my examinations as to his twenty-five years' Indian experiences with patience and intelligence. Also a Newcastle man (who introduced himself to me as one to whom my name was a household word! ahem!) travelling round world for his health. We discuss politics and northern affairs with zest. I have just been reading Cowen's last speech, which he lent me. Also an Australian doctor who has been ten years in practice, has been spending two years at hospitals of Vienna (where he says teaching splendid for students, but utmost brutality to patients), Paris, and Berlin; is now returning to practise as a specialist. Also Bailey (an engineer who has been twenty years in India, on various railways, as a contractor); has told me much as to native labour and habits—bright, clever little fellow. Also an Eurasian doctor (and wife); has been thirty years in practice in Calcutta, has two sons in Tasmania, where he is going to retire. Has son with him, much darker than father, though mother an European. . . . Also a China merchant who has told me much of China and Borneo. Besides, we have the new Bishop of Adelaide (an ex-Bradford cleric).

'. . . The day passes as to-day. . Up at 7.30 ; on deck

till 9, chattering to different people. Breakfast; then on
deck, chattering on New Zealand; and then with the
engineer. Then talk to Mrs. R. and Mrs. Professor.
Have short skirmish with the professor. Then the Spaniard
came and talked French to me, mostly jokes about Honman
(to whom he has taken a fancy and insists on talking
French to him, to H.'s utter confusion). . . . Then, to
make up, he gives H. a French lesson. Afternoon, a
group forms, and we have a general discussion (Honman,
incited thereto by jealousy, or envy and malice, declaring
that I lecture them all, and can be heard at the end
of the ship). Then a short read; then the Spaniard (by
the way he should be Argentine) and Honman come up,
and the Argentine gets off his *burlas* (jokes), and criticisms on the promenaders. H. bullies me about some
imaginary misdemeanours, and we find it's dinner time.
Evening: I write in saloon.

'*Tuesday, off St. George's Sound.*—Made some fresh
acquaintance. Bishop of Adelaide not at all a bad sort;
was telling me about a winter spent in Morocco for his
health, ten years ago. He speaks highly of Morocco. He
knows Middlesbro', and we did not collide any.

'Continue all right; though Honman says I ought to
spend next winter away, to which I demur strongly.—
Yours, dears, both,

'SID.'

The extremely favourable view Thomas here, as usual,
gives his family of his health is hardly borne out by his
faithful physician's letter from Colombo.

'We started from Calcutta,' Mr. Honman writes, 'under
rather unfavourable circumstances; for Sidney had caught
a cold at a dinner-party at Mr. H.'s on Sunday night, and
the next three or four days was suffering from a feverish

attack of bronchial catarrh. However, that has disappeared again. The symptoms of overwork have disappeared to a great extent. He sleeps better . . . and
looks less worried. The only thing that I am not satisfied
about is the condition of his lungs. The left has improved
considerably . . . but his right lung is still unsatisfactory.
. . . he has still a cough in the mornings, and (only
occasionally) during the day. Keep as much as possible
all work at home. *This is most important. Especially
any bearing upon Australian questions.* It will end in
interviews, negotiations, and business—never ending otherwise.'

Mr. Honman might well dread adventitious spurs to
energy. His patient, who draws above his own fancy
pictures of his pleasant 'loafing' existence, was in truth
constantly more than occupied with problems and questions
old and new, quite apart from pressure of the actual business and commercial affairs upon which he had embarked.
This latter class of work was, indeed, kept from him as
much as possible by his sister, who devoted herself to the
task of representing him, so far as she could, in his absence ;
but there were of course some matters which it was
absolutely necessary to submit to the decision of Thomas
himself. A source of anxious care at this period was the
nascent 'North-Eastern Steel Company' at Middlesbrough—mainly founded by Thomas. Unfortunately,
about this time a heavy depression set in in the iron
trade, and the new venture had to bear all the brunt.
'Sidney,' says his mother, 'always had perfect faith in its
future—especially managed as it was by Mr. Cooper. His
faith was amply justified in the result.'

The new problems he was perpetually engaged upon
were probably not so hurtful to his health, since in them

the element of anxiety was comparatively wanting. Some patents date from this time—one particularly for special steel sleepers for India. The utilisation of the slag formed in the Thomas-Gilchrist process was a matter which now and always occupied his mind.

CHAPTER XVIII

AUSTRALIA

On resuming the correspondence, we find Thomas on Australian soil.

To his Mother and Sister

' *March 6, '83: Melbourne; St. Kilda, West M.*—I wrote and cabled you on Saturday from Adelaide, where I landed and spent three hours on shore—making several calls and picking up some information. The city covers much ground, and is backed by hills about a mile behind—it being itself two or three miles from the sea. Everything, however, was baked brown, and an indescribable glowing sunshine pervaded all. There is every evidence of prosperity; but the place is not attractive, and one understands how great a refreshment the shadiness and dirt and air of long habitation of an *old* city must become to the dwellers in a new one. Arriving in the morning at nine, we left at five P.M., our passengers being diminished by twenty-four old ones, less a half-dozen new folk.

' A pleasant run close to coast (which is mostly sandy, but occasionally rocky cliffs); arriving inside Melbourne Heads at eight A.M. on 5th. Yesterday nothing happened on the way but a further closing up of acquaintanceship, pleasant talk with a New Zealand squatter and two other New Zealanders, who are all

enthusiastic about N. Z., and want me to go down there.
I think I must. I also want to see Tasmania; but how it
is all to be done I don't know. I think I shall have to
stop over till May, after all.

'Landing by boat, we came up to Melbourne by train,
and went at once to the Library,—a magnificent one,
where I revelled for two hours. They have, in same
building, a picture and sculpture gallery and museum.
There are some really fine pictures. I then called on the
man to whom P. gave me letter, and (in afternoon) came
out here, and settled into a pleasant little hotel facing
sea, where Mr. and Mrs. R. and two other fellow-passengers
turned up soon after, and we spent the evening together
very pleasantly. We do not go on board till to-morrow
at noon. The run up will only take thirty hours, so we
arrive 7th at Sydney, where I hope my "letter hunger"
will be satisfied. The suburbs of Melbourne bear every
evidence of prosperity, and some of the houses charming.
To-day H. gone to Hospital and races. R. gone to races.
I am going to make some calls and to the Library. A
bright clear day, but wind coldish. I feel first class, and
mean to stop so. . . .

'Shall return here before I leave Australia. A man
who joined at Adelaide came out in "Sobraon." He was
a special invalid, and is now quite well. He says, of
seventy passengers, sixty were invalids more or less; two
died on voyage. He says there is no doubt as to steam
being preferable. I thank my stars I did not go in her.
We had a number of affectionate partings yesterday.

'*March* 11 : *Sydney*.—Dearest Mother,—I wrote you
last from Melbourne, giving account of myself to date (by
the way, I have never yet missed a mail to you). That
morning Honman went to Hospital, and I into Melbourne
after seeing some people. . . . H. and I only next meeting

R

on going aboard (as he had been to Races and Theatre); he told me he had offer of *locum tenens* in a healthy place midway between Melbourne and Sydney for a month, which would give him a chance of seeing how he liked Australian practice, and yet rejoining me, if I liked it, or in fact my joining him, it being in district I am recommended by the Sydney doctor on board to go up to. As he was anxious to go, we arranged he should get his things off the ship and start at once, and let me know at Sydney if I should join him or go somewhere near.

'Going on board, we found only a third of our old number going on, though many of those who landed at Melbourne came to see us off. . . . Of my party there remained first and foremost my little New Zealanderin, Mrs. B., Mr. and Mrs. R., and the squatter millionaire. We formed a most pleasant set, and I made friends with various other passengers; so we were all like a family party. Starting at 1 P.M. Wednesday, we did not get on shore here till 9 A.M. Saturday, and I felt very sorry to break up even then. I had pleasant chats with young Victorian passenger, also with the Secretary of the Queensland Legislative Council, one of the oldest of Queensland's permanent officials. . . .

'At four on Friday we have our last "tea," the hostesses being Mrs. R. and Mrs. B. and the "Child"; guests, the ship's doctor and three officers, a nice, bright and cultivated old lady from Queensland, Miss T., the two R.N.'s just budding into uniform, L., and a few waifs. Such a laughing, childlike party as Lil would delight in. The Child decrees we are to have a final "race game," to which imbecile pastime we forthwith devote our whole energies, with the utmost gravity.

'Next morning we are all up at six, and enjoy the lovely view as we move slowly up the harbour to the wharf; the

R.'s and I go to same hotel, and we all disperse—the
" Child " being carried off to the new Premier's till she
sails for N. Z.   I rush to P. O. and get your three missing
Cape letters on paying a huge sum for accrued postage ;
then to D.'s, where more letters, but only one of later
date than those I got in Calcutta. . . .

    ' *Monday evening.*—On Saturday dined with L., who is
at another hotel, Mr. and Mrs. R., and a Col. H., an old
ex-army man of some family in Scotland ; knows everyone
here and has a lot of schemes.   We sat talking till nearly
midnight.

    ' Sunday I spent reading your letters and looking up
information about the Colonies. . . . Sydney streets are
largely traversed by tram lines, on which run large cars,
drawn by steam locos at a great rate.   They are an immense convenience and (astonishing to say) do not frighten
the horses.   I had two steam cars thundering down a hill
after my cab, their wheels almost touching ours, but the
horse did not move a nerve.   The park is large and beautiful, continues down to the harbour, and on Sunday was
full of well-dressed people, mostly work folk, I imagine,
quietly enjoying themselves.

    ' This morning have seen the Commissioner of Railways,
the ex-Premier, the present Premier, the Treasurer, the head
of Geological Survey, and a few others, and been generally
gassing around and acquiring piles of information.   I
have also had an interview with a female inventor and
patentee, who really knows something of what she spoke
of ; though she spoke of a good deal of which she knew
nothing.   I met the Premier at the Club.   Immediately
on introduction he ordered " five brandies " for myself
and himself, the late Secretary L., and two others.   This
solemn ceremony is colonial all over.

    ' They all abuse democracy and tell fearful stories of

the independence of working folk ; but I am inclined to
think things would not be half so well under any other
'cracy. The free libraries, accessibility of Ministers, cars,
parks, &c., are all democratic, and I *like* them. The
public buildings are very fine and convenient. The Free
Library (open all days, including Sundays) is alone worth
living in Sydney for. I spend a lot of time there. . . .

'I have a quiet day to-morrow, and expect to leave for
Wangaratta (Honman's place) on Wednesday or Thursday.
I am awfully good, and won't go out evenings, though
I should immensely like to. . . .

'Lil, dear, your letters are all that could be desired, we
must give you promotion. Try, darling, to understand
everything. You know why I want you to be posted in
everything. I boast no end of my little sister colleague.
Thank E. for her letter (amusing like herself). I hope
the Shipping Co. she has joined is Limited. Everything
depends on management. If Co. is *not* Limited, don't let
her put [in] more ; she had even better get out. Tell her
to read articles on shipping investments in "Whitehall
Review" of December and January. Weather just lovely :
hot in sun, cold in shade, and clear to distraction.

'If you still think it best, I am inclined to selling
house and carrying you both off, next winter, so as to run
no risks of relapses. I had almost forgotten to say I am
lusty and strong.

'*Wangaratta, Victoria : March* 18, '83.—Dearest Mother
and Lil,—Though I only wrote Tuesday, I won't let intervening mail go without writing.

'Wed. I went to R.'s to see their rooms, and then
with R. to see the Secretary of Works. In afternoon I
saw the "Child" off and made acquaintance with the
Premier's daughter, who came also to see her off; she a
bright girl, who, having been to Europe, pines to return

thither, as most girls seem to do. On Tuesday had been
to call on Mrs. B. who says I do not attend to social
duties. Laudable youth! . . .

'Had a comfortable berth in sleeping car and slept till
six, when we were traversing a dry, flat to undulating
land, covered with gum trees, mostly *barked* and dead,
giving a forlorn and desolate look. At 1 P.M. we came
to end of N. S. W. Railway and had four miles in coach,
crossing the river to the Victorian R. R. terminus. This is
a wine district; still arid and witheringly hot; but hills
and green trees and vineyards a relief. The river not of
much account now; but big bridges show what it is in
rainy season.

'At 4 P.M. got to Wangaratta; Honman at station to
meet me. Got a room at a nice little inn. His hospital
with a dozen beds (able to make up thirty) is only fifty
yards off; he sleeps there, but has his meals here. There
are five young fellows also boarding and sleeping here,
four bank clerks, and one the clerk to local justices, in fact,
pretty much what I was at Thames. The latter intelligent;
has told me a great deal about local conditions and politics.
. . . Excellent table, though simple.

'Honman gets all his exs. and a guinea a day, besides
some extras; thus he made extra 50s. yesterday. The
charge for visiting 5s. a mile. Thus, if patient is ten
miles from town they charge 50s., as was case yesterday,
when he and I drove to see a patient ten miles away, wife
of a small farmer, living in three-roomed house. It seemed
to me very hard lines that he should pay 50s.; but he did so,
and H. goes there again in a day or two. He is now off to
see a patient twelve miles in another direction. The hospital
is partly supported by Government (who give 900l. a year),
and balance by private subscriptions. The house-surgeon
gets 150l. a year and one room, but not, as I understand,

board.   He also takes as much private practice as he
likes.   Honman, you understand, is merely *locum tenens*
for a month. . . .   Honman drinks nothing and admits he
is the better for it.   He says the first days he was here
he was asked to have twenty drinks a day, but now no one
bothers him ; and I can see that he is respected for it.

'I have come here (though it is a dull place with nothing
near it that in any way interests me), because Honman
declares it is the most likely place to do me the maximum
of good,' and I thought you would like me to be near H.
or rather with him.   I therefore feel " awfully good " at
having banished myself from the attractions of Sydney
and not having gone to N. Z. or elsewhere. . I shall try
to hold out here for a month. . . . All right ; but oh! so
inexpressibly stiff after a two hours' ride on an aboriginal
quadruped.   I am going to get H. to rearticulate all my
joints.—Yours,

'S. G. T.'

'Wangaratta : March 22, 1883.
'(Thursday before Good Friday.)

'Dearests,—Life here is absolutely eventless, the only
thing happening being a rain-storm the evening before
last. . . .

'The magisterial clerk talks well enough.   He spent
three years in Queensland, by Gulf of Carpentaria, locating
a station, but got fever and scurvy and had to throw it up
and come back,—riding 1,000 miles to get a steamer back
to Victoria.

'My ride has not worn off yet.   I am even stiffer than at
Torquay.   I did not get as far as the hills, which are eight
miles away, and feel monotony of the everlasting gum
trees ; though these are by no means bad trees in their
way.

'The land here agricultural chiefly; but also largely cattle-raising; worth 3*l*. to 5*l*. an acre. One man has been here thirty years, has nice farm and six-roomed brick house, lives in plenty. Was a Bucks agricultural labourer at 12*s*. a week. The man who was to emigrate with him got frightened and stopped at home, and is still getting 12*s*. a week.

'Female servants get 10*s*. a week; said to be scarce, but the latter I fancy mistress' fault. The maid here does for five boarders (ourselves and two other family boarders); is always on hand, bright, quick, and smiling; has taken Honman under her wing, and dashes in with hot things for him whenever he comes in late. . . .

'The Athenæum here (free) is a glorious place. We have "Graphic," "Illustrated L. News," weekly edition of "Times," "Fortnightly," "Contemporary," "Westminster," "Cornhill," "Longman's"; besides Australian papers, periodicals, and a good library of good modern books. Have been enjoying "Other People's Children." Get "Realities of Irish Life," by Trench, one of the 6*d*. reprints—the best book I ever met on Ireland. . . . There is a strong anti-Irish feeling being got up here, particularly à propos of the Redmonds' visit.

'Fruit I am told grows here luxuriantly, though one doesn't notice it much. We get a reasonable amount—grapes 3*d*. and 6*d*. a lb.; in Sydney, even 8*d*.

'What [a] delicious, though impossible and irrational, book is "All Sorts and Conditions of Men!" Get cheap edition; it is worth having in the house, as a piece of dreamland. I don't do all novel reading; but (by dint of diligent study of Australian Gazetteers, handbooks, Mineral reports, &c.) am preparing to make myself *the* authority on Australian resources, so that I may "gas about" with effect.

' *Good Friday.—Specimen Day.*—Up at eight ; breakfast about till 9.30. Then over at Hospital with Honman; reading " Lancet," physiology, theology, &c., till lunch at 1 ; after which drove with H. to patient six miles off. Had chat with patient's husband : he took up the land (320 acres free) twenty years ago; farm now worth 1,300*l.* without stock ; has large family, all look not very healthy, mostly sore eyes, probably owing to flies and bad water. Untidy rambling low house ; plenty to eat; good farm machinery, reapers, chaff cutters, &c.; buggy. Would have been farm labourer at home. Orange growing here interesting, pleasant and profitable; but have to wait three years for fruit. There is a good deal of typhoid fever in outlying districts. On return, stop to chat with chemist, between whom and doctor there is the closest alliance. He comes from Totnes ; twenty-eight years in colony; free-thinker, intelligent, dogmatic. . . . Then look in at Athenæum (open every day in the year). Back to dinner. Honman called on six patients sixteen miles away. Honman proposes coming back with me, and then returning to Australia. His farmer patient to-day said, " New thing for us, a doctor who don't drink," and told how a predecessor came drunk, and severely injured him, performing an operation while drunk. The bank clerks here say the bank clerks in Melbourne are constantly drunk, say once a month or week. The young ladies of Australia are, I fancy, slightly American.

' By the way, dear child, you have still got to learn some Chemistry and work with me. I am absolutely brimming over with things that *demand* investigation ; the lines are already laid down and they *must* be investigated. I shall never have time by myself and you *must* help ; you can't tell what a glorious, entrancing, delightful occupation it will be, with rewards of the most magnificent description in reputation, work, benefits, and lucre.

' *Sunday evening.*—Yesterday and to-day idled away, reading, and good deal at Athenæum, and in open air. Been sixteen-mile drive with Honman to-day. Had long talks with several farmers and labourers; am becoming prodigiously learned on all agricultural matters. A man near here made 8,000*l.* this year out of fifteen acres of hops. Another, a carpenter, tells me he has been here sixteen years : earns 10*s.* a day; says working men can live cheaper here than at home; .meat 3½*d.* a lb., bread 3*d.* a loaf, flour 10*s.* per 100 lbs., clothing and groceries alone dearer. Education free; house rent cheap, and (land being cheap) can [live] out of town, have large garden, &c. This man, however, considers he is not one of the successful ones; says he could earn 7*s.* 6*d.* at home; his family middle class people living at Notting Hill.

' There being no poor laws, I fear there are many cases of hardship and even death of sick and old people. There are benevolent asylums; but difficult to get admission. In this little hospital the average of people brought in dying of starvation from remote parts is twenty a year ! An old man brought in last week, been lying in a field by the road starving for a week. He died without recovering consciousness.

' As illustration of colonial politics—at their worst, two incidents of last week: (a) A member, charged by the Premier in the House with saying in a speech to his constituents that he had seen thirty-five members of Parliament drunk, jauntily got up and said he *had* said so, knowing it was a lie, in order to influence votes in his favour. This is taken as a satisfactory apology, and an ex-Premier speaks of the M.P. in question, immediately after, as his "promising young friend." (b) A Cabinet Minister gets drunk at the Redmond Banquet, and makes an imbecile drunken speech; has in consequence to resign.

Petition to withdraw his resignation, as it was only a
trifling indiscretion—another M.P. in the house saying it
was cruel to take notice of such a thing, particularly as so
many leading public men were the biggest thieves on earth.

'*Wednesday.*—Long to be back to see you; otherwise
contented enough. If you see T. T., tell him I rely on his
trying the slag experiments thoroughly and having a
perfect slag process before I return. My heart is set on
this. I am sure I am on the right track. . . . Yours,

'S. G. T.'

'Wangaratta: April 2.

'Dearest Mother,—I really feel very cross and anxious
at receiving no news of you all since January 12. I know
you would not have left me so long, so conclude letters
have miscarried; but I am bothered just the same, as I
have got it into my head you may be ill. I got a whole
budget of papers last Thursday from Sydney, including
some you had sent to Cape. I so enjoyed reading even
the oldest. They were well selected too.

'My present plans include returning to Melbourne with
Honman, then to Sydney, then up country to quiet place
for a few days, then to New Zealand for ten days. Honman quite thinks to come back again. . . . He has been
out several nights, and has twenty cases in Hosp., all
more or less bad. A man brought in yesterday from fifty
miles away with a fractured thigh. We are here in the
heart of the bushranging country of a few years ago.
The sister of Kelly, *the* great bush-ranger, is now a patient
in the Hospital. I had a drive with Honman on Friday, a
ride on Thursday, and a longish walk yesterday; so I
know the country round well. Weather continues fine and
bright, though a good deal of rain has fallen during two
nights.

'I learn a good deal from the Magistrates' clerk of the business and social policy of the colony. I have been grinding up the resources of the various colonies from all sources, and it certainly seems to me that New Zealand is the best, New South Wales the second, or Queensland, if you have no regard to health considerations. In New Zealand good land, within thirty miles of a harbour, is still to be got at under 20s. an acre. Here the same land, only less fertile, costs 3l. and upwards, and in England 30l. and upwards. Am reading "Adam Bede;" a glorious book. This vegetating, I think, does me good, slow as it is.

'I am sorry to say I fear there is no prospect of starting Works in Australia, as I had hoped, so I have nothing to do but to loaf. Whenever you see T. T., tell him I am relying on his trying the slag experiments I sent him a list of; that I am sure the slag question is soluble in the way indicated, and that its present unsolved state is the great trouble of my life.

'By the way, I hope you were thoughtful enough to get three or four copies of "Cinderella." I go once a day to a place where they have been framed, to refresh myself by looking at her.

'I hope you are taking care of yourself; I hope too Lil has found some work of her own, in the direction of Besant's Angela or otherwise. . . .'

'Beratta, Victoria: April 8, 9 A.M.

'(In hotel verandah, in a very comfortable chair.)

'Dearest of Mothers,—This is intended to be a birthday letter, and I hope the P. O. will arrange for delivery accordingly. That I wish you ever and ever so many happy returns of the day, and that you may continue in your special way to grow younger and younger, as your offspring grows venerabler and venerabler—all goes without saying. Your second sight, or affectionate intuition, ought to be telling you all the time how much I am always thinking of my facetious little mother. I sometimes think of setting up a special shrine, on your plan (with travelling lamp attached) for your and Lil's photos. I am feeling peculiarly bright and brisk; the receipt at Melbourne of your letters of Jan. 19 and Feb. 21 (which only reached me April 6 after a month's blank) was an immense relief. . . .

'As I wrote you last week, we went down to Melbourne; only came from there last evening. I do *not* care for Melbourne; though there is much life and animation, still the country round is flat and uninteresting, and it does not do after Sydney. The hotels and buildings, Public Library, Museum, &c., are all finer than in Sydney, and it is much ahead in population. I had planned to go over from Melbourne to Tasmania, which I much wanted a glimpse of, and I also much wanted to see an Iron Works there for which I have interesting views; but I got your letter on the morning of starting, and (as Honman seemed to think my going to a colder place injudicious) I gave it up with much groaning and tribulation. Now I call that an exhibition of gorgeous abnegation of my own (better) judgment. . . . It is much colder in Melbourne than at Wangaratta. Thermo. about 65° in shade, which *I* call cold. . . . I may stop here two days; then to Fitzroy for a day or two; get to Sydney about Thursday, stop three or four days to find out some of the people I have introductions to, and then up country again quietly. Start for home about mid May; it is uncertain whether by U.S. or by the Orient, or Messageries. Honman returns with me.

'All the Australian towns seem just like one another. Buildings mostly one-storied, some brick, some wood;

balconies and verandahs wherever practicable. Wide roads. Country round often looks like wilderness, or a ragged English park; generally a river about six feet wide, with a bridge sixty feet wide, to provide for floods. Bright blue sky, clear air, bright sun, now often cool wind. Two banks, public library, first-class school. Lots of stores, and every fourth house an hotel or drink shop. People here seem religious; in Wangaratta (with about 600 people) a Roman Catholic Church, Church of England, Wesleyan, Presbyterian, Independent. R.C. has schools, and excommunicates all who send children to State Schools, which here are free, and I am told very good. Very loyal and patriotic too. At present all papers are abusing Ireland and the Irish, and circulating and believing ridiculous atrocity stories. Railways all State; indifferently managed and undermanned. Porters remember they are Government officials, and act accordingly. . . .'

'Sydney: April 12.

'Dearest Ones,—Your home news may seem trivial to you, but it is delicious to receive out here. I will certainly be back before July 20.

'Now for a spell of gossip. I spent last Sunday very quietly as I wrote you, mostly on verandah. A curious incident was [the] passing of a small "selectors'" funeral to R.C. Cemetery. First a hearse; then about a dozen buggies, carts, and traps of various kinds, all full of decent poorish country folk; then thirty-six men riding two and two on horseback, some smart, some shabby, some ragged, most dirty, some with a bit of black tied on, mostly without. It was curiously impressive, motley as it was. The deceased, it seems, belonged to what they call the "Kelly Crowd," i.e., the friends of Kelly, the notorious bushranger, who lived between here and Wangaratta. Did I tell you his pretty young sister was in Hospital, a patient of Honman's, from anæmia? H. says simply hunted about and worried into severe illness by the police.

'In evening I found that a man I had talked to in morning, and taken for a com. traveller, was the M.P. for the district. Picked up from him and others local information. Land round Beratta very good; most of it worth 3l. to 4l. an acre on average. You hear constantly of English farm labourers now farming 500 or 1,000 acres of their own. Wages for agricultural labourer 20s. a week and board, but said to be hard to get.

'At 11 A.M. Wednesday started by train back through Wangaratta to Wodonga; then three miles coaching and through Albury by train through Wagga, Macdonald's nearest station, to Mittogovey, seventy miles from Sydney, and on top of hills 2,000 feet high. Got there late; knocked up landlord; got in.

'Next morning found it a curious big public, with (as usual here) several boarders. We all mealed together. We sat down to dinner, host, hostess, two daughters of about twenty (to whom I devoted myself); a Chinaman; an Irish shopman; a railway porter; a storekeeper (ex-gold-digger in Transvaal, bit of a carpenter and doctor also, and quite a character, became quite a chum of mine); a hawker and itinerant quack. This last been all over world; entertained me with account of a trip from 'Frisco to New York knife and scissor grinding. Three or four odd lots, diggers, labourers, &c., and an aboriginal. In my two days I conversed more or less with all. Spent morning in talking to landlord, an ex-policeman, ex-auctioneer, ex-storekeeper, &c., and going over the abandoned Fitzroy Ironworks, which I enjoyed. Afternoon went for a walk; was introduced to leading citizens and storekeepers. They had a general idea that I was either emissary of Rothschilds', an impecunious digger, or a lunatic.

' Next morning at 7 an intelligent quarryman came with two horses to take me to see a geological phenomenon which they told me I couldn't find by myself. After a time we struck into bush and rode for some way up and down hills, among the forest. . . . My horse shied at the first Australian bear I have seen, not bigger than a big poodle, climbing up a tree. My guide then began riding down a precipice, and I made my will, strapped myself on to my horse, and requested that animal to do with me what he would. The result was the quadruped proceeded to walk up and down vertical walls of a few hundred feet high (with superb trees growing at the bottom) for some four miles, occasionally having a quiet jump across a mountain or river, and I enjoyed it very much. I think, however, my guide did not think I was such a good talker as he had been led to expect, as I found fastening myself on required considerable attention.

' The scenery in those precipitous rock gorges really very fine and enjoyable. The phenomena, which were of a carboniferous character, proved very interesting, and I rehoved and restrapped myself on to my charger and trotted gaily back, leaving it as before to my friend the horse to say whether he should proceed on his hind legs, or his forelegs, or his tail, exclusively or otherwise. Generally speaking, he would coil his tail round a tree and then drop down on his hind legs to the next valley. Anyhow, he understood the country, and we got up a showy gallop when we got within sight of the hotel. The young women removed my remains from the saddle, and I felt good for dinner. My companion was very intelligent, and I collected mines of notes which my executors will believe are mutilated cuneiform inscriptions.

'Left Mittogovey by 2 P.M., not a whit the worse for my ride, which I had actually enjoyed immensely. Found two men in carriage, the one a colliery owner, the other a merchant; plunged into discussions. Lots more information, exchanged cards, spread myself out.

'Got to Sydney 6 P.M.; came here (better than other hotel); found here P. (my Newcastle "Paramatta" friend) and Mr. W. and his sister (of Liverpool), who came out in "Paramatta" for their health, intending to stop here only a few weeks. They both look worlds better, and are going to New Zealand and Tasmania before they return. He talks of settling here. After dinner chat with Miss W., and joined by Miss T., who also here with her people. They also from "Paramatta." Find two others from P. also here. Weather bright and pleasant. Next day get my delicious big budget of letters; revel in it. Call on big firm [of] merchants here. G.'s step-son pleasant, sharp Scotchman; gives me some information I want. . . . Called also yesterday on Watson, ex-Colonial Treasurer, pleasant bright man, Scotch; interesting short talk; had some trouble to avoid invitation to dinner, which I do not want. Read some time at Royal Society's library. To-day (Saturday) have made some calls, had a photo taken (which is hideous in the extreme), to please you. . . . Land here at present is at a fabulous price; had gone up five-fold in five years. . . . Went to the Picture Gallery, a small good collection, and Botanical Gardens and Domain Park, coming down to harbour, hilly ground very well laid out, making a lovely park. . . .'

'April 18.

'Dearest Mother,—I have little to add to my hugely long letter posted on the 16th per Orient S.S. Monday I called on one of the ex-Ministers, a Jewish merchant. . . .

I spent some time at Library, wrote letters, &c. Yesterday called on one of my fellow-passengers; then drove
to University, saw Professor S. (another shipmate); his
class as yet only five; very busy. They have allowed him
to spend over 1,000l. in specimens and apparatus, and
give him all in buildings &c. that he wants. They intend
to have a first-class Medical School. Then called again on
professor of chemistry, who showed me round and thawed.
. . . Called on Sydney Jones yesterday. He has big
practice; very pleasant. He examined me, recommended
me not to stop in England next winter. Honman says
same. . . . S. Jones comes home same time as I do for a
two years' holiday. He advises me to go on hills, so I
go up to Lithgow to-morrow. I may then go up to
Brisbane, which he also recommends me to do. We had
heavy showers yesterday and to-day, but bright sun meantime.

'Thursday, 19th, Noon.—Just got yours of 9th. I wish
I were worth one-third of the thought you give me. Your
letters make me feel quite ashamed always of not being
worthy of your goodness. Lovely weather. Sitting
writing in verandah. Honman goes with me to-morrow
into the hills.—Ever yours.'

'Sydney: Saturday Night, April 21, 1883.

'Dearest Mother,—On Wednesday afternoon took a
trip up the Paramatta River for the greater part of its
course, and round the harbour to Paramatta, one of the
oldest towns. Started at 1 P.M.; arrived at 3; back here
by 5. The whole way a panorama of pretty scenes, wooded
knolls, and bold rocks. For first three or four miles from
Sydney large numbers of suburban villas and villages; these
grow fewer as you go further. The harbour lovely to a
degree; sites overlooking it now selling at enormous rates.

S

'Had pleasant chat with old boy who had been twenty-eight years in Melbourne; was an official on Victorian Rails. Being now entitled to retire, was speculating if he could live in Europe without the sun. Evening, chat on balcony with various hotel acquaintance. By the way, my first appearance in antipodean journalism (a short editorial article) was an anonymous letter of mine to "Sydney Morning Herald" on behalf of the caged monkeys of Botanical Gardens, which I had to interfere with roughs for ill-using on Sunday.

'Thursday got your letters of 9th in morning; had to scurry to reply by mail leaving two hours after. In afternoon went on board the "'Frisco" mail boat to see Mrs. W. and her brother off on their way to New Zealand. I was tempted to go down to New Zealand too; but they say it is too cold at present, so I have resigned the hope of seeing New Zealand . . . this time. Archbishop Vaughan (Catholic Metropolitan of Australia) sailed for Europe by same boat. The Catholics had been holding farewell meetings and addresses for several days, and had given him 3,000l. for pocket money, and now crowded steamer, and had lines to small steamers which ·were thronged with people (some thousands) to accompany him down harbour. It was a curious sight. He (a fine-looking man six feet high) stood on top deck, with gold chain and eccentric (Archbishop's) costume, waving hand as they cheered and waved handkerchiefs, till ship out of sight. A splendid vessel. I hope to sail by the next month's boat, if we can get cabins.

'Did I tell you of going over ironmongery store of L., one of our "Paramatta" fellow-passengers? . . . A vast place, steam engines, tools, machinery, ironmongery, china, glass, furnishing, natural gems, wire plates, &c. &c. Turns over 500,000l. a year, and (I suppose) nets 40,000l. or so.

'Friday at 9 off by train to Lithgow, crossing on way the Blue Mountains, 3,000 feet high, by zigzags. Superb views for three hours; highest point Mount Victoria, a great tourist's place. Talked with Scotch clergyman now in Sydney; very intelligent; says no poverty here, except from drink or improvidence. We talked much together on poverty, its remedies, workmen, &c. Very liberal, enlightened man. Asked me to call and see him.

'At 3 got to Lithgow, in valley, 600 feet below Mount Victoria; pretty, but collieries and an Ironworks. Hotel moderate; hobnobbed with other guests a com. traveller from Belfast, Ireland—intelligent; came here partly for health; much better.

'This morning went over Works, formed opinion, got lots of information. Manager bad lungs; says Sydney suits him best; says labour costs twice that of English labour; interested, became great friends.

'Left at 3 P.M. for here; at station chatted with man of sixty-five, a selector in hills, born here, brought up thirteen children, who are well educated; ." is not lern't himself, but knows things." Has house and bit of land; still has to work, "but has his victuals and his bed, and don't see he'd be better off if he was Lord Chief Justice of New South Wales, as his school-mate, Sir F. Martin, is." (N.B. Find he has iron ore on or near his bit of land.) "Has been a pioneer; rough times, seen men speared by blacks, may have shot some blacks; opines he has; but won't be sure if you saw a man who might spear you, you would [not] think it safest to shoot him. Father lived to be ninety-six; expects to do the same." Had difficulty in getting into hotel, being all full; at last got half a room with my sick com. friend from Lithgow. Have been talking Irish politics to him. Bathurst big place; lots of churches, bishops, and institutions; is on the high-lying plain at foot of Blue Mountains.

'*Sunday.*—Vile wet day; fortunately comfortably lodged, bar bedroom. Had fire last night and this morning, and quite enjoy it, half wood half coal. I am quite childishly looking forward to seeing you both; am wearying of wandering, though there is much I enjoy. I have not yet got any papers by last Suez mail, so do not know anything about time of I. and S.I. spring meeting. I hope to get Suez papers when I return to Sydney. Fear there is no chance of picking up a lovely girl! . . . I am pretty clear they won't come out to be picked up—by me. I hope to get more letters here; and a budget at Palace Hotel, 'Frisco, if I cable you I return that way. . . . I am still inclined to think that London for next winter would be injudicious. I have no desire to run risks, to entail more banishment. If I. & S. Meeting in Middlesbrough, you must make your long promised trip to York with me, staying there or going on as you prefer. I know Lil will prefer staying at York, so we will leave her there anyhow. . . .

'I go to a place thirty miles from here by rail to-morrow; said to be very pleasant; stop there a few days, then back to Sydney for a few days; then probably to Brisbane, as it is getting hot here; back to Sydney and so homewards. Hurrah!—Lovingly yours.'

'Sydney.

'Dearest,—Just starting with Honman for Brisbane by his advice, so as to get a spell of warm weather before leaving by 17th May, on which berths booked. Shall cable you if nothing occurs to change plans. It is rather ruinous dashing about so much; but I am become reckless, in Colonial fashion, of expeditions. . . .'

'Brisbane: May 1, 1883.

'Dearest Ones,—Here I am in a new colony and new life again. . . . Brisbane boat close quarters after the P. and O. Enjoyed much the steaming up the harbour, in praise of which one can't say too much. Had beautiful passage, close to high rocky coast all the time. A coast range of hills, unfortunately, between coast and inland. The line of coast far prettier than the line of English coast on an average; but very little settled, land not being good; several good harbours. Passengers uninteresting as a whole. One had been ten years on cattle station; well educated; said he began with too little capital, and has always regretted it. Says you ought to have at least 3,000l. to 4,000l., and that if you have 8,000l. or more, you ought to make 18 to 20 per cent. In this all agree. Cattle worth 8l. each, fat sheep 10s. to 13s. Got much warmer weather on Sunday. On Monday at 2 got into Moreton Bay, and soon entered river; fine winding stream, banks high one side, generally low on other; luxuriant vegetation, pretty houses, mills, &c. at intervals.

'Brisbane, about thirty miles up, looks like a compromise between a huge country village and a big city. Fine buildings everywhere, with trees and gardens sprinkled about. Landed at 6 P.M.; nice hotel, all on ground floor, somewhat Indian; found our fellow-passenger whom we met at Sydney at hotel; chat, dinner, to bed betimes as usual. To-day like a hot English summer day, everything bright and pleasant. Going out for a walk. We have taken passage by New York route; start for 'Frisco on 17th, arrive at 'Frisco 14th June. I feel like a schoolboy at prospect of getting home and seeing you. Got your two birthday charming letters on Saturday, just as leaving for here. . . . I grieve at not being able to stop at New York, but Honman, I think, advises not.

'*Tuesday morning, May 2.*—Spent yesterday loafing
in reading-room, Botanic Gardens, and about. Weather
delicious, though perhaps air a trifle too moist. Land in
Brisbane has increased four times in value in last six
years. Best frontages now sell for over 266*l.* a foot, i.e. for
a frontage of 100 feet the price is over 20,000*l.* Thirty
years ago you could have bought the whole city for a fourth
of this sum. There is a vast inland country, say as big as
England, France, and Spain, which is now found to be
rich cattle and sheep land, and coast land is already
enormously used for sugar. One " squatter " here, worth
three millions, is said to live as he did when he had a few
hundreds, spending much of his time passing from one of
his stations to another, sleeping on ground, feeding on
" dampers " &c., never having new clothes, and never
spending on anything beyond necessaries. Millionaires
are absurdly abundant here, and men talk of square miles
as we do of acres. I have had many offers of leases of
1,000 square miles, the rents of which are often only 10*s.*
a mile, while good-will fetches scores of thousands. One
station recently (but this freehold) sold for over 300,000*l.*
. . . The more I see of Brisbane the finer does its
situation seem . . . on the bend of a fine river with high
rocky banks, and wooded hills as a background. Hurrah!
Shall see you in ten weeks. Love to all.—Yours ever.

'*Friday, May 4.*—Dearest Ones,—Though I only posted
to you on Tuesday I will send this line as an Orient S.S.
is leaving. Tuesday, reading-room and gardens; the
latter very pretty, on a peninsula, surrounded by river, to
which they slope. Cricket and lawn tennis in full force,
notwithstanding the heat. Hotel very comfortable. Made
acquaintance with a Scotchman who has recently come out
to look after business of a great Glasgow thread house.
He gets 1,000*l.* a year and expenses, all out of reels of

cotton &c. Yesterday same routine; Honman spent evening with leading doctor here; there are twenty-three doctors here for 30,000 people. Is no opening except up country. Read Sullivan's "New Zealand;" very good. Still lovely weather. Go down to Sydney in a few days. Start on 17th. Hurrah!!'

We have now for a long time been following Thomas's admirable letters from Australia. We will presently give Mr. Honman's health report, which, as usual, corrects Sidney's own too optimistic view: but let us interrupt the Australian letters at this point, to relate what the Iron and Steel Institute was contemporaneously doing in England, to pay honour to the young inventor. In 1873 Sir Henry Bessemer had founded, under the auspices of the Institute, a gold medal, to be awarded annually by that body, to persons distinguished by their inventions. or services in promoting the manufacture of iron or steel. The Council of the Institute in this year, 1883, resolved to award two Bessemer gold medals—one to Thomas, and the other to Mr. Snelus, whose connection with the basic process we have noticed above.[1]

The Institute held its spring meeting on May 9 in London. Thomas was, of course, in Australia, and, at his mother's request, the actual presentation of his medal was deferred to the autumn meeting. Thomas, it will be remembered, had during the preceding year been elected a member of the Council of the Institute—succeeding Sir James Ramsden, who himself succeeded the ill-fated Lord Frederick Cavendish, as one of the vice-presidents.

A day or two before this meeting Thomas was beginning the following last letter home from Australia.

[1] *Ante*, p. 135,

To his Mother and Sister.

'Bellevue Family & Squatters' Hotel, George Street, Brisbane
(opposite Botanical Gardens and Parliament House, Brisbane):
'May 7, 1883.

'Dearest Ones,—I hope to follow within a fortnight of this, but I certainly shan't get home before July 15, possibly not till 20th. . . . I wrote you on Friday last. Friday afternoon I spent in Gardens, and calling on the Clerk of Executive Council, who showed us over Parliament Houses; fine buildings, but Parliament not now sitting. Rather a rowdy lot, I gather, have got in lately.

'Saturday.—Calls, reading-room, Gardens, &c.

'Sunday.—Dined with a merchant to whom I had an introduction from a business friend; bachelor, new house on river, two miles out, pretty view. Banker dined with us; pleasant talk; they had both been round trip by America, Japan, &c. All say New Zealand has finest scenery in the world. We go down to Sydney to-morrow; raining to-day.

'*May* 12, '83.—Got yours of February 24 only to-day, as it was not addressed by Brindisi. . . . *Revenons à notre* diary. On Tuesday last we started for Sydney per steamer, my merchant friend seeing us off. Had a beautiful sail down the river and along the coast; chatted with passengers on Northern Queensland and Queensland politics (on which I am proficient), land laws, &c., wool, and beef, and democracy. We sail within half a mile to a mile of the coast nearly all the way, there being a range of hills twice, coast and good inland country. Next day at noon wind began freshening till it got so fresh that at teatime I and Honman felt that eating was a morbid carnal craving of unregenerate man, which ought to be suppressed. It finally got so remarkably fresh that we concluded to

seek the retirement which a small cabin with closed ports
and all the hatches battened down gives so sweetly, and I
began offering fabulous rewards to anyone who would
drown me out of hand. As everyone, however, was
occupied in a private service of groaning on his own
account (H. included), no volunteer handy. We finally
got to Sydney on Thursday evening, slightly the worse for
wear. One lady passenger was delirious, and very ill.
Honman stopped with her on board for some hours.

'Friday, went and talked to the Premier about some
ideas of mine. It was, unfortunately, deputation day, and
(as the Premier is now holding two offices, Colonial Secre-
taryship, and Minister of Works) I had the opportunity
of seeing the poor man chevied about all over the building
by hungry packs of subsidy seekers.

'Sunday, May 13.—Yesterday made a call or two in
the morning. Met Professor S., who made me promise to
go up to his house to-day. In afternoon young S. came
to hotel; interviewed me at great length on European
politics, literature, &c. . . Honman and I go along lovingly.
He proposes to come back with me to see America, though
of course there is no necessity for it. I am quite rejoiced
at the prospect of getting nearer home from Thursday
next.

'To-day Club in afternoon; then with Prof. L. to tea
at Prof. S. Latter just got into eight-roomed, single-storey
house, rent 150l. . . .

'Monday.—Called about; raining all day. Evening
dined at Club with Professor L. Old School of Mines man;
had next bench to Percy. Has 1,000l. a year as professor.
Lives at Club, where it costs he says 250l. a year. Is an
F.R.S. and clever . . . was . . . very nice to me. Sydney
merchant dined with us told us many things. . . . Says
miners of a concern he is director of earn 50s. to 70s. a

week. Were earning 20*s.* to 25*s.* at home. This morning, Tuesday, saw Railway Commissioner. Profitable
chat. . . .
    ' *Thursday,* 17, *Noon.*—Been interviewing Premier and
Treasurer. Very busy, having great fun bullying ministers.
Lovely day ; feel *very* well, as I could for next five years.
Honman and I go on board at 2. Been farewell visiting.
Flourish of trumpets. Hurrah ! Shall see you in two
months. Take care of yourselves. Mind, I am first class
in health.'

    ' First class in health !' Such is Thomas's last message
from Australian soil to his ' dearest ones ' at home. Let
us turn to Mr. Honman's reports, sent from time to time,
during the two months' sojourn in the Southern Continent.
    From the ' Paramatta,' Mr. Honman had written :—

    ' I am sorry I cannot say that his lungs have improved
much.'

    From Adelaide on March 4, 1883, he wrote :—

    ' I have examined Sidney's lungs this morning ; the
left is greatly improved, the right has improved sufficiently
to give satisfaction.'

    From Wangaratta, the stay at which up-country place
Thomas has described above, comes really the first
reassuring news. On March 14 Mr. Honman writes :—

    ' Sidney has been improving gradually since my last
letter, and I can at last report some decided improvements.
His left lung is better and his right is improved to a great
extent. His general health is also better. I have been
stopping here and at Melbourne for the last three or four

days, and to-morrow Sidney joins me again. It is a very good place, and more suitable for him than any we have yet been at.'

When Thomas himself gets to Wangaratta the intelligence is still better. On March 26 Mr. Honman writes to Mrs. Thomas :—

'I have had Sidney here again, and am so far satisfied with his condition. Our climate here is perfection.

' . . . He will still persist in working out some scheme of an Ironworks here. . . . It seems impossible either to prevent him working or talking. . . . I have been able to take him some long rides in the buggy through the bush, and he is always ready to act as a Jehu and pilot the horses along. The drives are delicious here, in fresh warm air, through miles of bush—the "bush" consisting of big red gum trees and other aromatic smelling trees. The air is so clear that hills that are ten miles away appear to be but half-an-hour's walk. . . . This seems to me the best climate we have yet reached, and the healthiest, I fancy. . . . Sidney's chest has not improved much; but his general health *has* improved. . . . I don't think we can do better than here.'

Thomas's mother and sister were so much struck with the good reports of Wangaratta that they wrote entreating Sidney to remain there, and offering to wind up affairs in England and join him in Australia. Thomas talked sometimes, as we shall see immediately, of reverting to his early love—medicine, and qualifying for a physician's career. Knowing that he would never consent to a life of idleness, and that a strong counter-attraction would be required to distract him from metallurgical problems remaining to be solved, the solutions of which could only be satisfactorily procured in Europe, it was suggested that he might, in partnership with Mr. Honman or otherwise, become a doctor in Australia. Unhappily, the letter containing these proposals only reached the antipodes after Sidney's departure therefrom. Perhaps, despite the little improvement ever really manifest in the lung, his life might yet have been saved had he received this letter, acted upon it, and settled at Wangaratta. It is sufficiently useless to speculate upon such might-have-beens. As it was, the letter was returned to the senders months afterwards, when the dear doomed one was already entering into the Valley of the Shadow.

On April 10, 1883, Mr. Honman wrote from Melbourne:—

'Sidney has been stopping with me at Wangaratta, and it has done him a great deal of good. Your letter to him arrived very opportunely; he had determined to go to Tasmania against my wish or permission. . . . However . . . I have sent him North, where we shall be constantly heading for now.

'I wish I could tell you his lungs were highly satisfactory. I cannot indeed do this. His right still remains the same; his left is better, but for the emphysema. I have endeavoured to persuade him, although it would be painful to you, that he should not go back till the next summer; but I am afraid he will not consent to this. I said I should be no tie to him, because I should set up here, and he could enter into partnership with me; he always declares that he is the best doctor of the two; and I have proposed another plan—that he remain here, and I go home.'

The effect of all Mr. Honman's letters is the same. The

general health improves, but the lung trouble never disappears. He writes from Sydney on May 16, at the very moment almost that Thomas is describing his 'first-class health.' 'Cough a little troublesome . . . The months on board ship ought to improve him more.'

CHAPTER XIX

HOMEWARD BOUND

LET us quit for a little the slow process of measurement of the advancing steps of Death, and revert to Thomas's own correspondence, brimful as it is of life.

To his Mother and Sister.

'SS. "Zealandia"—off New Zealand : May 21, '83.

' Dearest Ones,—As I wrote you per P. and O., mail left Sydney Thursday at 3 P.M.   Tuesday and Wednesday spent in interviewing Premier and Treasurer, who mildly complained that I treated them in a most unceremonious "stand-and-deliver" fashion, but showed by their action that it was the right line.   They had a Cabinet Council on me, and were greatly disturbed at my audacity, and wound up with saying that they were favourably impressed, but wanted time to consider.   All this showed much of the interior working of colonial politics, and kept me quite amused. . . . Sydney Harbour looked its best in the bright sun as we steamed out.   We had had a week's rainy and cold weather, so appreciated the bright sun the more.

' The vessel a good one, with admirable arrangements for the passengers,—the saloon and stateroom being forward of the engines.   There are eighty or ninety passengers in the saloon ; thirty more join at Auckland.   I have chatted with twenty or thirty of the crew.   Among them are our

South American shipmate from Calcutta to Australia: a
pressman and ex-Victorian M.P., going to report for his
paper on the United States, with whom I chat much:
Speaker of Victorian Assembly, who lost his arm in heading
miners' revolt against authorities thirty years ago: a
Brisbane doctor; a Queensland sugar-grower: a South
Australian wine-grower: two or three health travellers:
two young squatters ; four girls, and eight or nine married
women: two Roman Catholic priests, and a Victorian
Anglican cleric.

'May 28, '83.—We arrived at Auckland late at night
this day week. I went on shore before breakfast next
morning and took train across the island, to see the only
Ironworks in New Zealand. Particularly interesting, as
being trial of a new process. Saw manager &c. Returned
to Auckland. Made a call; got some useful information on
several subjects. It was by this time raining hard, so took
a cab back to ship, and we steamed away at 2 P.M. with
twenty new passengers. The glimpse of New Zealand I
had was pleasant. It is greener even than England.
Abundant vegetation and picturesque rocky coasts and
hills. . . .

'There are three doctors among passengers. . . . There
is also a Belfast man, who has for some years been
wintering in Australia, who has ideas, and with whom I
discuss politics sometimes; and a Sydney man from
Canada, who is bright and intelligent. I am making him
read "Progress and Poverty." Gambling on the "run"
occupies two-thirds of the time of two-thirds of the
passengers. I, of course, keep out of it. . . . They have
had a dance and a concert and games. P. (my South
American acquaintance) is very popular. The other night
he ordered champagne all round to drink to Argentine
Republic, on anniversary of its formation He came to

me to write an English version of the French speech he
proposed to make. I rewrote an English version of a
gorgeous description, and coached him how to deliver it;
but at the last moment his courage failed him, and he
asked me to read it, which I did; so finding I have not
quite lost my voice. It amused me to hear P. con-
gratulated on the English of his speech and its periods—
congratulations which he received with great modesty and
satisfaction, and an occasional smile and bow. . . .

'*June* 4, '83.—I feel good, when I think I am now
only six weeks from home, at most. Our voyage to
Honolulu, where we arrived at noon on Sunday last, quite
uneventful. . . . Have discovered another bright fellow,
a young Cornishman, who is partner in a large New
Zealand business [of] the London house which he entered
as a clerk nine years ago. He is not only clear-headed on
business, but has read, can talk, has thought, speaks French
and German, plays the piano, and draws clever caricatures.

'We were at Honolulu from noon to midnight on Sunday, June 3. The Island, of which Honolulu is the
chief town, is volcanic and rather picturesque; vegetation nearly tropical, sugar-cane chief crop. The natives
rather fine-looking, identical with Maoris of New Zealand;
women, however, get stout and coarse-looking early.
We landed at one, and I sent Honman for a drive. I
(strolling round) picked up a young fellow, a cabinet-
maker from San 'Frisco, who showed me round till 6 P.M.,
all over the town and surrounding country. My guide
proved very conversable and well-informed, and posted me
thoroughly in Hawaian matters. He (though only twenty-
one) was making about 7*l*. a week. The country is
particularly "run" by Americans, who control the chief
political posts and the bulk of the business. The half-

caste girls are singularly good-looking, with clear, brilliant
olive to white complexions. The King is given to drink,
but is otherwise a good constitutional sovereign, that is,
does nothing, draws an enormous salary, and gets into
debt. There is a large Chinese population living entirely
to itself. I went through and through the Chinese
quarter, which is densely crowded. Here, as elsewhere,
they propose shutting the Chinese out. Labourers there
now get 6s. to 12s. a day : artisans 12s. to 20s. Rent is
dear—20s. a week for a four-room house ; but food cheap
enough. We took fifty passengers on board for 'Frisco, so
are crowded to a degree.

'*Sunday, June* 10.—Thank Heaven, we have but one
day more before we see shore and get letters. I am more
tired of this trip than of any of the others, and weary for
the land. The past week has been coldish, and sufficiently
rough to prevent being on deck, so we have been nearly
confined to the smoking-room and saloon, both of which
stuffy. . . . I hope to see you all before July 17. Have
been very well all trip, though still obliged to be careful.
Honman seems to be for coming home. . . . Don't be
making engagements for July or August. I want to see
as much of you as I can.

'*Tuesday morning, June* 12, '83.—We arrived in 'Frisco
last night. Just on shore : all well.'

To his Sister

'Palace Hotel, San Francisco : June 13, 1883.

'Dearest Lil,—The Palace Hotel is truly palatial.

'Like city well enough. Weather bright and sunny ;
coldish winds. We leave to-morrow [for] Laramie City.
I hate delaying a day, but at same time want to gather
any information that may be useful to N.E.S. Co. on way.
I don't see way to getting home before 15th. I got

T

"Ironmonger," and of course much interested in report
of meeting. . . . I walked several miles yesterday without being tired. . . . If any of my fellow-travellers
call before I return, you will, of course, do the right
thing, and tell them when I return.

'Lovingly yours, dears,
'SID.'

But while Thomas was writing thus cheerfully home
of 'walking several miles,' Mr. Honman was describing
the true state of affairs, viz., that he was in a 'dangerous'
condition, and quite unfit to remain in England.

On June 14, 1883, the latter writes from the Palace
Hotel :—

'I have had the opportunity of examining Sidney off
ship and in a quiet place. His right lung is still dangerous
and gives me a great deal of anxiety. It is absolutely
imperative that he should leave England immediately the
more important business matters have been settled, or else
entirely drop business matters for the autumn and winter
months (and this latter, I presume, would be an impossi-
bility if he were to remain in England). I wish that his
condition had been free from everything to cause anxiety.
Had it been so, I should have remained in Australia. . . .
Sidney has been walking about all day in great spirits.
We have lovely weather, but with a fearfully cold wind at
night. . . .'

Here this long correspondence practically ceases.

Thomas, now nearing England, no longer writes
voluminous epistles, but confines himself to short letters
and post-cards. We give some of these in their
order :—

To his Mother and Sister

' *Laramie City, June* 17.—Arrived here all right. . . .
All way very comfortable ; had ten of " Zealandia " passengers with us.   Some very fine scenery, but most
monotonous plains. . . . Get to New York about July 1.

'S. G. T.'

'Pittsburg: June 24, Tuesday.

' Dearest Folk,—Arrived here last night and got your
letters (with delight as usual) of 31st and 5th. . . . I
haven't so far found United States at all too hot.   In fact,
I can stand any heat. . . . I was kindly received at
Cleveland ; driven about, taken over Works &c.   Saw G.,
who sent messages to you.   He is earnest and innocent as
ever.   At Chicago saw M. ; had F. to dinner, who drove
us about and took us to Club &c.   Taken to Cleveland in
state in Dunlow car.   Had rather hot ride here, starting
at three and arriving nine.   Had G. and his chief to early
dinner with me before I left.   This American part is
proving very costly ; have been twelve days in United
States, and have spent over 60*l.*, *besides* railway tickets.
Have just met two Liverpool men who are stopping here,
going round the world the other way.   Am going to call
on Mr. Tom Carnegie.   Will now only write you post-
cards, or shall have nothing left to tell you when I come
back.   I must be in London till the end of August, or
nearly so ; can't be back now till 18th or 19th.   I wish
we could have got earlier passage.   Love to all.   Tell A.
I expect him to be M.R.C.S. when I return.—Ever yours,
best ones,   ' SID.'

' *Pittsburg, June* 26, '83.—Spent very pleasant evening
on Sunday with Tom Carnegie. . . .   This morning been to
Works, quite leisurely and easily, declining to exert myself

T 2

any. Pleasant reception everywhere, much kindness.
Get on to Philadelphia to-morrow ; easy travelling. Shall
have a week in New York, which I worry over; am so
anxious for return.

'*Philadelphia*, *June* 28, '83.—Left Pittsburg Tuesday,
after seeing a little more quietly. Yesterday drove down
to Steel Works at Harrisburg, where working Basic.
Very kindly received. Came on to-day to Philadelphia;
lovely day; taking it very easily. Get to New York on
Monday. I wish I could sail at once.

'*New York*, *July* 6, '83.—Just another line to say all
well. Weather still very hot ; shall be glad to be on ocean
again. Everything improving since last here; colossal
buildings everywhere, both office-blocks, hotels, and apart-
ment houses. This hotel has been beautified enormously,
less high,—art restaurant and ultra high art Bar, with
good oil paintings, statues, bridges, antiques, &c., painted
windows and—iced drinks.

'Honman and I spent 4th July, when New York is
shut and deserted (except by youthful fiends letting off
crackers), mostly in Central Park. In evening to theatre.

'Yesterday, calls ; dined at seaside. We go on board
to-night. Sail to-morrow. Unfortunately I shan't get home
till three days after this [arrives], as our "Nevada" is a
very slow boat.'

CHAPTER XX

A SAD HOME COMING AND A FLIGHT SOUTH

'His voyage from the States,' says his sister, 'was not made under satisfactory conditions. In his haste to get home he had wired to a friend in New York to secure berths in the first ship. This happened to be the "Nevada," a vessel chiefly used to convey Mormon parties to the States. It was old, slow, and badly ventilated.

'Letters calculated to worry him reached him at Queenstown. The very day of his arrival at Tedworth Square visitors, requests for appointments, business of all kinds, began to pour in upon him. It was quite evident to us at once that his health could not withstand the strain, and we made despairing attempts to keep work from him, attempts mostly made in vain. It was well nigh impossible to check his activity and eagerness.'

Not alone had he to deal with the many questions constantly arising in connection with his various patents, with the development of the basic process, and with the progress of the North Eastern Steel Co.'s Works—some of which questions had necessarily been reserved for his consideration upon his return—but the very travels primarily undertaken in search of health had produced a new crop of plans and problems to be worked out. From every country he had visited, he had brought back a mass of figures and economic statistics, together with general information of all kinds. He had occupied himself with the special circumstances affecting iron and steel in South Africa. He had entered into lengthy negotiations with the Government of India, with a view to purchasing from them certain ironworks, collieries, &c., his object being to establish (or rather to re-establish) steel manufacture in the peninsula. This was a matter which he had very much at heart, not only from a commercial standpoint, but also as a right and proper effort to give back to Hindostan an ancient industry which the British Raj had destroyed. As will have been seen, he constantly dwells upon the subject in his letters. For Australia there were schemes for the foundation of fresh colonial steel works.

These were no idle phantasies of an imaginative inventor. It must be remembered that, from the first ' blast,' Thomas had had the sole legal and financial conduct of all matters connected with the basic process. The rapid and absolute success of that process is the best possible tribute to his practical ability and clear grasp of realities. That success was not won without some sharp legal contests; above all, many delicate and difficult negotiations were needed to secure the fruits of discovery. The very important North Eastern Steel Works, started at Middlesbrough to work the process, owe their existence chiefly to Sidney's initiative.

Beyond all these things, the question of the utilisation of the ' slag ' produced in the basic process was a problem which from this time, for the few remaining months of his short life, more and more dominated Thomas's never quiescent mind. Of that problem and its thoroughly successful solution we will speak presently.

The pressure of work and the harassing business interviews, soon destroyed whatever good the voyage round the world had wrought, and after a fortnight of London, it became very clear that town must be quitted at once, and England itself at the first opportunity. In the first days of August, Thomas and his sister went down to stay at the White Hart, Sevenoaks Common, leaving their mother to wind up matters in Tedworth Square in preparation for a long absence from British soil.

'He and I,' says his sister, 'thus set out once more on the health quest, this time together. Our month at Sevenoaks was happy in its way (happy since we were once more together), although it gave me too many grievous proofs of his frailty of health, and too much of that anxiety of heart which seems most overwhelming when one realises that cherished hopes have been disappointed. We worked together, and in the intervals of work sauntered along the country lanes or sat in the old-fashioned inn garden. Many kind friends came down to see us. The last Directors' Meeting of the North Eastern Steel Company which Sidney was ever able to attend was held specially at the White Hart,—the other Directors thoughtfully travelling south to meet my brother, inasmuch as he was quite unable to go to Middlesbrough to meet them.'

Thomas wrote a letter from Sevenoaks to his old chemical teacher, part of which we reproduce:—

To Mr. Chaloner

'Sevenoaks: August 28, 1883.

'Dear Chaloner,—I should have answered yours of Saturday before but for a tremendous influx of business (from which I still suffer) keeping me hard at it all day, while we have two Directors' Meetings for to-morrow. . . .

'The fact is I have thrown my health and *everything else* into the basic business, and it is possible I may not see the harvest myself. But we shall see. Thanks *very*

much for taking so much trouble about Algiers. We shall
be able to do with Murray, which I have ordered. I
shudder to think of the ten volumes.

' . . . We shall sleep at Dover; probably spend
Sunday there.

'We have almost, not quite, settled to go Saturday, if
I can finish off business by then, so can hardly hope to
see you. . . . In haste, yours very truly,

'S. G. THOMAS.'

So long as any physical power remained, even reasonable rest was impossible to Thomas. 'Sidney,' says his
mother, 'instead of resting, was interviewing at the White
Hart his cousin Mr. Gilchrist, his secretary and chemical
clerk Mr. Twynam, numerous friends, anxious to say good-
bye. His brother, Dr. Llewellyn Thomas, was quite
overcome at discovering the rapid change for the worse
which had set in since Sidney's return to England. The
change made little difference in my boy's ardour for work.
"Mother," he would constantly tell me, "I have so much
to do." Much time was necessarily occupied by writing
business instructions to those he left behind him in London
and Middlesbrough. He had a long day with his lawyer,
arranging all his affairs.

'I joined my children at Sevenoaks on August 25.
Sidney, although unfit for it, insisted on driving to meet me
at the station. I saw at once that the two or three weeks
which had passed had left him weaker even than he had
been in London. We drove the two miles to the White
Hart sadly and almost in silence.'

After some anxious days of waiting, the little party
began to journey southwards, taking advantage (on
September 8) of the first fine day to cross the Channel
and gain Paris. After much study of the advantages and

disadvantages of various Mediterranean health resorts, Algiers had been pitched upon as upon the whole the best place to winter in,—Cairo (whither Thomas had wished to go) being shut to him by the cholera, which was then raging there.

'We stayed only long enough in Paris,' says Sidney's sister, ' to make some necessary financial arrangements and travelled on to Marseilles, breaking our journey at Lyons. Boats do not go every day to Algiers, and some days had to be spent in hot, dusty, noisy Marseilles.'

The turmoil characteristic of the great southern sea-port tried Thomas (now, in truth, an invalid) much, and he became alarmingly worse. He was removed to an hotel some three miles along the seashore, at the end of the Prado, and grew better again. 'We waited here,' says his sister, 'happily enough, save perhaps for the mosquitoes, out of which, even, Sid managed to extract fun,—describing his skirmishes with them in grandiloquent and Homeric terms, and trying various languages in which to summon me to aid in a conflict with them,—finally declaring that, though they understood French and English, German was too much for them, so that they did not know when we plotted their extermination in that tongue.'

CHAPTER XXI

A WINTER IN ALGIERS

On September 22 the little family got themselves on board
the Algiers packet. 'We were two nights at sea,' says
his mother ; ' Sidney better, as he always was at sea. We
landed at 6 A.M. on the 24th. The juxtaposition of
Eastern and French civilisation much impressed my son,
as it impresses everyone. Before 7 A.M. we had driven
into the courtyard of the Hotel Kirsch, where we were
received by sleepy servants, evidently surprised at European
health-seekers coming to Africa so early in the autumn.
We soon discovered that we were the very first guests of
the season, full three weeks too early. The ground was
still parched from the summer heats and all vegetation
had withered away. The sun shone with a constant hard
glare and the deep blue sky remained from morning till
night without the shadow of a cloud to veil its brightness.
Sidney became very ill from the fatigue of the journey and
from the prostrating heat. The English physician had not
yet arrived for the winter, and we sent for a kind French
doctor (an Alsatian, whose own excellent health had been
built up by the Algerian climate). He evidently thought
my poor boy in a very bad way; but—after one or two
visits—he said that his courage and mental force gave him
a chance. On this foundation we raised great hopes.

' I even now think that, if we could have kept his

mind quite at rest, he might have rallied, but this was
impossible. Letters poured in, causes for anxiety arose,
and no effort or persuasion could induce Sidney to "let the
world slide as they did in the golden days." Even during
the three weeks of summer heat, he would insist on
driving out almost daily to look for a house. Fortunately
we consulted the excellent British Consul, Colonel Playfair,
and he pointed out to us that most of the pretty houses
we saw, and were pleased with, were badly drained. So
for the present we stayed on at the Hotel Kirsch.'

A part of Thomas's correspondence with England
referred to the presentation of the Bessemer Medal, a
presentation which had been, as we have seen, postponed
from the spring meeting of the Iron and Steel Institute.
Thomas was quite unable, of course, to be at Middlesbrough
to receive the honour. He wrote, however, a letter of
thanks to the President, as characteristic, in its generous
tribute to others and in its self-effacement, as anything he
ever penned.

'It would be difficult,' he says, ' for me to insist too
strongly on how greatly we are indebted for the success
the basic process has now attained to the unwearied
exertions, the conspicuous energy and ability, of my
colleague, Mr. Gilchrist, whom I regard as no less my
associate in the acceptance of this medal than he was in
the sometimes anxious days of which this is the outcome.
I am sure, too, that he and I are agreed in saying that the
present position of dephosphorisation has been only
rendered possible by the frank, generous, and unreserved
co-operation of Mr. Richards. As an instance of the
effect of free discussion of metallurgical theories and
experience which this Institute especially promotes, it may
be interesting to note that, while in the autumn of 1877
there was, so far as I know, no public record of even any

successful experiment tending to show that phosphorus
could be removed in the Bessemer or Siemens process,
for the present month of September 1883 the make of
dephosphorised Bessemer and Siemens steel is between
60,000 and 70,000 tons.' [1]

By a happy departure from usage, the actual ceremony
of ' presentation ' of the medal was, in this instance,
performed by Sir Henry Bessemer himself.

We resume Thomas's mother's narrative :—

' After our three weeks of drought, clouds suddenly
gathered, and we had such a downpour of rain as two of
us, at least, had never seen before. After that the weather
was perfect and everything grew into delicious life.
About this time an invalid Irish gentleman arrived at the
hotel, who became a great friend of Sidney's. He had
lived many years in Paris, and had come thence to Algiers
seeking renewed health. Many discussions did he and
Sidney have on Ireland and her needs, politics in general,
or on the prospects of the Algerian colony. We spent
four months and a half in the Hotel Kirsch,—Sidney
fluctuating much, but always steadily working, and fighting
against his disease. We passed our time entirely together,
—he, his sister, and myself.

' Friends gradually gathered round us (Sidney made
friends wherever he went), and, as we were still buoyed
up by hope, the time passed not unhappily, in spite of
terrible dreads. Sidney was always cheerful and even
vivacious, save when unusually weak. He would eagerly
join in the conversation at our end of the table d'hôte,
bringing his varied knowledge and acquired experience to
bear on current topics. Once a week or so, when Sidney
felt well enough, we would drive into Algiers and sit in

[1] For the present output of Basic steel see *post*, ' Conclusion'; *cf. ante*,
p. 159.

the great *place*, watching the different nationalities and
gaining peeps at Arab life.'

In the following letters Thomas gives some glimpse of
his Algerian impressions :—

To Mr. Chaloner [1]

'Hotel Kirsch, Mustapha, Alger: October 4, 1883.

'My dear C.,—After seeing you when you last so kindly
enlivened me at Sevenoaks, I had some days of being very
much indeed under the weather. Lil said I talked to you
too much, which I denied as the *causa mali*. Once started,
took very slow stages,—sleeping one night at Dover, two
Paris, three Lyons (which is bright interesting town), and
stopping ten days at Marseilles (where at last I found it
decently warm). The last town looks very flourishing and
busy, is well-ordered, and from the sea looks magnificent;
but for smells it beats Paris at 2 A.M.

'Crossed here. The town of Algiers looks well from
the sea, with high green trees all round it; it is built on
slopes and steeps. Here, we are two miles from the town
and some 700 feet or more above it, looking on the bay.
We came here direct, and shall stop for some months
anyhow. Town very interesting; mixture of new French
town and slip of Arabia and the Patriarchs. Camels and
tramcars; mosques and chapels; Arabs and Parisians;
steam-engine and hand-pounding of wheat. The natives
and immigrants are unanimous only in fleecing the stranger.
Hope to benefit. At present find it too cold at 70°.

'Yours,

'S. G. T.'

To Miss Burton

'Dear Bess,—Many happy returns of the day, and no
more returns of any failing in health ! These are the best

[1] On a post-card.

wishes I can wish for you. We were so very glad to hear that your holiday did you so much good, and that you had returned quite bright and well.

'Whether I shall ever get round enough to enjoy a *real* holiday is dubious; but meantime I *ought* to be enjoying this wondrously sunny place,—which is, for the rest, interesting enough otherwise, if I could get about more.

'We live pretty much entirely in our own rooms. I have plenty to think and write about: so we are not altogether dull.—Yours,

'S. G. T.'

Innumerable letters on financial and chemical matters of course continued to arrive and need reply. Thomas's correspondence alone would have been sufficient to tax the energies of a vigorous man; but the brain of this invalid was constantly occupied with engrossing thoughts of all kinds, and with fresh projects quite unconnected with current business. Truly the 'aspiring spirit' 'o'er-informed its tenement of clay.'

'Among his ideas at this time,' says his sister, who was ever his indefatigable helper, 'was a plan for an improved type-writer, in which he sought to interest his old friend and teacher Mr. Chaloner, who was to help him with it in England. Many were the trials we made in the Hotel Kirsch drawing-room of the relative speed with which I could strike the piano keys with my finger or with rods of varying lengths, and many are the sketches he made of his improvements,—sketches which remain to testify to a portion of the work still left for him to do, but which he was prevented from accomplishing.'

The type-writer project is spoken of in the following letter :—

To Mr. Chaloner

'Hotel Kirsch, Mustapha Sup., Alger : November 20, 1883.

'My dear Chaloner,—Many thanks for your card. . . . I should have written you long since but for the extreme weariness I generally feel after getting through with the little necessary writing of the day : add to which the life here is eventless absolutely. The weather is, after all, the only thing to talk about—and that is certainly superb. Sun, sun, and again sun! though (alas) we are now degenerating into 50° F. at night, and have had three wet (and so fire-needing) days ; but to-day it has been 100° in the sun again, and I breathe once more, *literally*.

' Have not been up to any foot rambles now ; but what we have seen in driving of the country is pretty and fertile —orange and olive trees ; vineyards and all sorts and kinds of vegetables ; flowers in bloom (roses, geraniums) everywhere. I could wish it 10° hotter all round ; but one can't get everything to suit.

' We are stopping at a sort of compromise between a hotel and a boarding-house ; pleasant enough,—the host and hostess kind and obliging. . . . I am . . . satisfied to be left to my books, we having four tiny rooms to ourselves.

'I have several fresh things on hand which may, or not, fructify. Among other things, the improved typewriter [1] I have seen the way to for some years past, and spoken to you of before, has turned up again. I think a type-writer could be sold to write 15 or 20 per cent. quicker than present, and at half the price. If so, it means considerable money. . . .

'I grieve very much over the steel trade. Prices are

[1] It will be remembered, of course, that all this was written in 1883. Probably type-writers have been radically improved since then. These matters are only inserted here as illustrative of Thomas's character and bent of mind.

lower than ever; but on the whole I doubt if it will hurt
our prospects, from a patent point of view, ultimately.
But meantime we are squeezed some getting along on the
Continent!

'Lily insists on my leaving off.—Yours,

'S. G. T.'

A month or two later Thomas, in his regular correspondence with Mr. Chaloner, recurs to the type-writer in a
long letter, too long and too technical for profitable
reproduction. However, we may perhaps insert here a draft
Memorandum on the matter which was enclosed therein :—

'*Memo.—re Type-writer.*—The only two type-writers
in practical use are the Remington and the Hall.

'These defective as follows: (*a*) Price: Remington
costs 13*l.* to 25*l.*; Hall, I believe, 7*l.* 7*s.* (*b*) Both fatigue
the wrist, elbow, and shoulder joints; while the Hall also
cramps two fingers. In both the whole hand and arm
have to be moved to reach and depress a key. The action
of the Hall is especially fatiguing and cramping; the
striking of each key necessitating considerable muscular
force. In both the eyes are strained to catch the type-marks
of the keys—the Hall notably very defective in this
respect, the effort rapidly producing head-ache and exhaustion. (*c*) In the Remington the large number of
complicated jointed levers exposes the machine to
frequent disarrangement, and it is very hard for the user
to repair it.

'My object is to produce a machine which shall not
cost more than 50*s.* to manufacture wholesale; that will
require a minimum movement of the hand or fingers and
no muscular exertion, combined with simplicity and the
possibility of much greater rapidity than can be attained
in the present machines.

'To attain this :—

'1. I use type set radially or circumferentially on a wheel or quadrant.

'2. I cause a given type to be brought into-striking position by raising or depressing a key by electro-magnetic instead of muscular force.

'3. I make contact with the keys, and so establish the current which brings the type into place by means of a short rod or light hammer which enables the comparatively slow motion of the fingers which guided to be translated into a very rapid motion of the striking end of the rod. It can be shown by experiment that twice the rapidity of key striking can be obtained by the use of this hammer that is obtained by the unaided finger or hand.

'4. The necessity of striking exactly on a particular key is obviated by the use of angular guides into which prolongations of the keys fall, and which guide the type-bearer into its exact position. This also enables a much larger type-indicating board to be used, which can be placed in any convenient position, so that the eyes are not strained.

'It would appear that the idea of setting the type radially on a circular surface or wheel has been suggested before, and it is not proposed to claim this, or to claim any special method of moving the paper upward or forward, this being done either by a rack and pinion, or a screw and a ratchet.

'In my proposal two or three bichromate cells would supply the power. Compressed air &c. might be proposed as alternatives, but would be less convenient and efficient.

'The only items of cost in such a machine would be : (a) the type-wheel, which could be of ebonite with the type cast on it. This, with the keys and attachment for

U

bringing the type into striking position, would cost less
than 10s. The (b) framework and paper, advancing
screws and ratchets, might cost 10s. more. Case (c) &c.
5s. Battery (d) and electro-magnet with attachments,
say 12s. *Items:* (e) 8s. ; or (say) 45s. in all.

'The points are—use of electro-magnet in place
of muscles; long hammer in place of moving arm and
hand; use of angular guides for bringing types into
exact position.

'If the thing could be sold for 4l. and enable an
ordinary person to write sixty words a minute, I would
contract for 100,000.'

Thomas was, however, diverted from his type-writer
by the more pressing interest of slag-utilisation.

On February 7, 1884, the family removed from the
Hotel Kirsch to Bir-el-Droodj, an English-built house
near the village of El Biar, which is situate on very high
ground, three or four miles from the city of Algiers.
Here Thomas was able to have a laboratory of his own,
and could work at various haunting problems, above all at
that special problem of the utilisation of basic 'slag,'
which, as we have said above,[1] was becoming more and
more the dominant question of all to him.

'The slag matter,' says his sister, 'tormented him.
How right he was as to the capital importance of this
question will be seen when I state that, in 1889, 700,000
tons of basic (or "Thomas") slag were produced (containing thirty-six per cent. of phosphate of lime), and
that most of this immense quantity of slag was used as
a fertiliser, being applied directly to the land as a
manure.

'In the winter of 1883–84, this valuable product was

[1] *Ante,* p. 278.

looked upon in England as so much mere troublesome
rubbish, to be got rid of somehow—by stacking on waste
ground—or even by taking it out to sea in barges and
there depositing it. In Germany things were more advanced. The mode of utilising slag, which has eventually
proved commercially successful, viz., grinding it to a fine
powder, had already been tried on the oolitic ores of Ikert,
at Peine, by Herr Hoyermann. About 1880 that gentleman had applied the grinding treatment to the puddle
slag produced at the Peine Works. On the great success
of the Thomas process in Germany, Herr Meyer, Chairman
of the Peine Works, pointed out to Hoyermann the greater
richness in phosphorus of the "Thomas slag." Such
slag was, therefore, substituted for puddle slag with
thoroughly satisfactory results. In the winter of 1882–83,
what is now known as "Thomas phosphate powder" was
first tried on the land in Germany as a manure, and in
November 1883 Herren Hoyermann and Meyer were able
to report to the German Royal Agricultural Society most
excellent effects from its use.

'These details, however, were not at the time known
out of Germany. Sidney, for all that, had long had a
very practical belief in the future of the basic slag.
Already, early in 1882, he had induced a few other
metallurgists to join with him in purchasing and stacking
this "waste product," as it was then supposed to be,
relying upon his ability ultimately to turn it to account.
From Algeria he wrote to Mr. Gilchrist, strongly expressing his views on the slag question, and putting them in
what must then have seemed a very paradoxical form
(although the paradox has already to a great extent proved
true) :

"'However laughable you may consider the notion, I

am convinced that eventually, *taking cost of production into consideration*, the steel will be the by-product, and phosphorus the main product." '

To Mr. Chaloner, it may be observed, Thomas uses very similar language. On February 15, 1884, he writes to him from Bir-el-Droodj :—

'I should have written to you long ago had I not been so seedy that I have had to reserve all my strength in the writing way for pressing regular business, and the development of certain theoretical views which may or may not turn out to have considerable practical consequences. . . .

'I have recently patented provisionally certain ideas of mine connected with the alkali trade &c. They have been verified to a considerable extent ; but (for my complete specification) I want to have the result of certain other experiments, which will be pretty numerous, and require considerable care and some partial analyses.'

And on the 29th of the same month he writes to the same correspondent :—

'My idea, which I have already patented under five heads in separate patents, is this. I propose to make steel as a by-product in a new alkali trade. . . .

'You see, according to my old principle, I have taken a big contract, and I intend to take it through. There is a big stake at the end.'

We resume Thomas's sister's narrative :—

'He also suggested to Mr. Gilchrist a series of fresh experiments on slag utilisation, which he wished him to undertake; but his cousin did not desire any fresh work, and declined to help in this direction.

'Sidney therefore enlisted the services of Mr. Twynam, his valued assistant, who had (as had also Mr. Aldred) carried on experiments for him for some years. In a short time three other chemists were also working at "slag," upon lines laid down by my brother.

'Two distinct processes were tried at this time. By the one it was sought to extract the phosphorus from the slag by the use of acids. By the other (which became of absorbing interest to Sidney) the object was to so treat the iron, while in the process of conversion, that the phosphorus in the slag should be deposited in the form of soluble phosphates, which would need no treatment to render them immediately agriculturally useful.

'Sidney would often talk to me in Algeria, not only of the necessity of utilising the slag in order to further improve the position of the basic process, but also of the benefits to agriculture which would accrue from making useful such a vast mass of material. He often quoted the saying about the benefit to mankind of making two blades of grass grow where one grew before, and described the fields of corn which would ripen in the future upon "basic slag."

'A very competent authority thus writes of the importance attached by my brother to the slag as early as March 1884 :—

'"I may say that Mr. Thomas was the man in connection with the North Eastern Steel Company who first appreciated the important part in basic steel manufacture that basic slag was destined to play. In March 1884 we had some negotiations with a large firm who wanted to buy our slag over a term of years." (At this time, the "waste product" might have been reasonably

considered to be advantageously disposed of on any terms.)
" Whilst we were negotiating with them, we received a
letter from Mr. Thomas, in which he said that, so important did he consider the slag question to us, that if we did
enter into an arrangement with anyone over a long term,
he would have to consider whether he would not sell out
his interest in the company. This letter influenced us
greatly, and I am quite clear that, at one time, he was the
only one of us all who appreciated the value of the slag."

'It is now evident that, had the slag been sold forward
at the low price it would then have fetched, the company
would, to say the least of it, have been seriously
hampered.

'In this same month of March 1884, Sidney, wishing to
superintend the slag experiments himself, arranged for
Mr. Twynam (his able assistant above mentioned) to come
out to Algiers. The whole of the miniature "plant"
needed had to be imported from England, and there were
journeys down to the quay to arrange for its landing and
conveyance to our villa upon the plateau of El Biar. It
was, however, when the strange packages had been safely
carried up by the bare-legged Arabs, and the whole apparatus reared in the court-yard (looking oddly out of
place amid its surroundings) that our difficulties began.

'For fuel we had to use wood, charcoal, and coal,—
there being no gas. For the blast there was a "foot-
blower" which needed a human foot to move it, and we
were all needed for other posts. I sallied forth to El Biar
village to procure a man to work the blower, and soon
engaged an Arab willing to take the payment per
hour Sidney offered, which was sufficiently high. However, when our Arab presented himself and was shown the
work he had to do (merely to work with his foot the
bellows supplying the miniature converter) he shook his

head gravely, and departed without a word of explanation. After this we had many applicants "to see the machine ;" but having been shown it, they either left silently, or else, when the fire being lighted and the "blast" starting the sparks began to fly, they took an early opportunity to glide away. We found that they considered the apparatus an "infernal machine" at the very least. In the end, however, we found a young Arab who took everything that happened with the greatest and most imperturbable coolness. Sparks might fly, molten metal splutter when poured, this Ishmaelite at any rate evinced no emotion of any kind, but went calmly on with his work, only pausing to change from one foot to the other. Afternoon tea was always brought out to us in the courtyard, and "our Arab" (as we called him) would accept a cup with the same gracious dignity with which he worked the blower. When, at the end of two or three hours of experimenting, he retired, he would gather a sweet-scented flower or two (always with permission), stick his nosegay behind his ear, gravely salute, and leave.

'We must have made a strange scene in that Arab courtyard. On two sides of it stood our English-built but quite Algerian villa, on the third an old Arab house and "loggia" joined to the villa, the fourth side was open, save for a low wall, beneath which the hill sloped down to a little valley running towards the sea. On the ground floor of the older Arab house Sidney had established his small laboratory. In the midst of the courtyard, with the "loggia" as background, stood a palm, with pansies at its feet, and a great Roman vessel of earthenware, dug up in the vicinity, beside it. To one side was the little Bessemer converter. Sidney would sit in a delightfully sheltered invalid chair (lent by kind friends) and thence direct operations,—now and then dashing down the books and

papers of which his chair was always full and sallying forth
to lend a hand to Mr. Twynam at the pot, to be forthwith
driven back. Meanwhile " our Arab," with crossed arms,
red fez, bare legs, and white garments, gravely worked the
bellows with his foot.

'The experiments were continued with varying success,
hampered a good deal as they were by the difficulty,—
either of getting up sufficient heat, or of repairing any
little accident to the apparatus. Many apparently insurmountable obstacles were overcome by Sidney's inability
to feel himself beaten, and fertility of resource.'

In June he wrote to Mr. Gilchrist:—

'I wish I could convince you that our one hope of
reducing costs is in slag, as I am sure it is. Remember
the phosphorus is more valuable than the iron in pig—only
we are too stupid to turn it to account properly.'

Alas, amid all this eagerness to follow fresh paths of
discovery, Thomas was not growing better—rather, the
fatal lung disease was strengthening its hold upon him.
His sister gives two illustrations of his persistence in
attempting to ignore weakness. The first has reference to
his sensitiveness to anything in the shape of cruelty to
the lower animals, an example of which has already been
noted (ante, p. 258). The present writer well remembers
his growing almost angry in argument (a rare thing indeed
with him) because the said writer defended vivisection by
some possibly too sweeping assertions as to morality not
applying to our dealings with brutes.

'He was constantly,' says his sister, 'interfering on
behalf of dumb creatures. One day on our way town-
wards, I parted from him to make some inquiry. On my
return I found Sidney breathless and exhausted, and found

from the friend with him that he had interfered to prevent a driver (who had called in a soldier to assist him) from belabouring an unfortunate overdriven horse who found it hard to toil up the steep hill. He had succeeded in stopping the ill-treatment, and had sent the driver back for another horse to help draw the load; but his success was at the cost of great exhaustion to himself, partly from his indignation, partly from the effort he made to keep his indignation in some check. All strong emotions exhausted him, and the more because of the self-repression he always exercised. The strain only showed in the lines of his face and the added pallor of his complexion.'

The other illustration of his readiness to plunge, ill (indeed dying) as he was (although the latter condition he did not yet realise), into physical exertion is of a different kind.

'We found a difficulty,' says his sister again, 'in getting satisfactory copies of letters, having brought no copying press with us. He declared that "if he had two boards, a rope, and a pole, or plank, he could rig up a gorgeous press." I thought no more of this declaration. Next day, however, I met a procession up our leafy lane, consisting of Sidney and a friend, carrying a plank some four or five feet long between them—Sidney so scant of breath as to be scarcely able to speak. I, of course, assailed him with reproaches, when he humbly explained that he had slipped out after *déjeuner* to the village, and had procured a satisfactory plank for his press from the French carpenter. It being, however, the siesta hour, he could find no one to carry it, and had consequently shouldered it himself. On his way he had met his friend, who was sufficiently astonished to see the invalid in such guise, and who had naturally insisted on bearing a portion of the burden. The copying press was forthwith constructed, and remained in use till the end of our Algerian sojourn.'

There were many visitors and callers at Bir-el-Droodj. Thomas would especially enjoy long talks with Mr. John Bell, with Colonel Playfair, the British Consul, and with Mr. Boys, the Anglican chaplain. Lady Macfarren (then also an Algerian sojourner, and next-door neighbour to the Thomas family) would regularly come in to play to Sidney, and her visits were sources of great delight to the invalid, who was always passionately fond of music.

When not in his chair in the courtyard, superintending the 'blows' of the little converter, Thomas would (if the day were warm) spend his time in a hammock in the garden, reading, writing, meditating. 'He would lie,' says his mother, 'in his hammock, a pile of books and papers by his side, absorbed in thoughts, calculations, or diagrams. One of us would be always with him, although he might not speak for hours. If we left him for a few moments, he would soon grow restless and would be gazing up the garden for us as we returned.'

On a perfectly still day he would sometimes drive down into Algiers with his mother and sister, in a little pony chaise. 'One beautiful Sunday in May, I remember, especially,' says his mother, 'we drove through the city and up to the church of *Notre Dame d'Afrique*, built on a high hill overlooking the sea, to hear the fine service and see the procession from the church doors to the edge of the hill—a procession in which the priests offered prayers for those at sea, and a hymn was sung. The scene, beneath the African sun and upon the shores of the blue Mediterranean Sea, was a very impressive one. Sidney was tired, but took no harm.'

Thomas still kept up a correspondence with the staff at the Thames Police Court, especially with one who had been really a friend of his, although in what would be called a subordinate position, R., the gaoler of the Court,

who in 1884 was still at his post, although over eighty
years of age. No man ever lived with less of class feeling
than Sidney Thomas; for him what has been called the
'class war' was as non-existent as, under present arrangements, it can be for anybody. He met R., whom he liked
and respected, as he met everybody else, on a footing of
absolute equality. The following letter from Mr. Lushington was, it will be seen, written in consequence of a letter
from Thomas to R., and shows the feelings with which the
whole staff at Thames regarded Sidney :—

Mr. Lushington to S. G. Thomas

Thames Police Court : May 30, 1884.

'Dear Mr. Thomas,—R. showed me a letter from you
a few days since, from which I gather that you are wisely
staying in your Southern quarters till the summer has
really set in, and then only coming to the north of the
Pyrenees, or some such climate, not trusting yourself in
this treacherous east-windy England. I fear the winter
has not been a very favourable one for you as far as
weather goes. I hope you don't let the chemical amuse-
ments which you mention to R. exaggerate themselves into
any such prolonged occupation as to affect your health.
You have made such a mark upon the world that you have
every right to try and enjoy your success as happily and
easily as the misfortune of your weak health will permit you.

'Do you happen to have read Nasmyth's "Autobiography?" It is to me one of the most delightful books I
have seen for a long time. Probably you have; but I
mention it as a possible amusement in case it should not
have come across you.

'You will be glad to hear that old R. appears to me as
vigorous as I have seen him for several years, and he has
had a very good winter. Most of your acquaintances here

are gone. You would find a great change if you could see
the amount of work here now compared to what it used
to be. The Thames is getting the reputation of being
one of the light Courts. . . . With all wishes for health
and happiness in all ways,—I am, yours very truly,

'F. LUSHINGTON.'

Thomas's health was in a far more serious condition
than probably Mr. Lushington supposed. Possibly, as we
have said above, if he could have abstained from work,
and, above all, if he could have been kept free from the
anxiety of many business complications which had followed upon his achieved success, his life might have been
somewhat at least prolonged; although the disease of the
lung had probably by this time progressed too far to make
final recovery in any case likely; but rest from further
labour was quite outside the limits of possibility to one of
his mental constitution, and freedom from anxiety was not
vouchsafed to him. His mother in her diary repeatedly
notes the arrival of worrying letters and consequent
aggravation of distressing symptoms.

'Our good French doctor, M. Bruch,' says Mrs.
Thomas, 'would stroke his head and say, "Keep him
quiet," "Keep him from writing and thinking;" but this
was just what could not be done.

'His physical state fluctuated much; but on the
whole, even in Algeria, he grew feebler. Drives fatigued
him more and more, and he more and more rarely ventured
away from his hammock or his hooded invalid chair. His
patient endurance was wonderful; never through all his
sharp attacks of chest pain or through all the prostrating
exhaustion which followed, did he cease to be our dear
thoughtful companion, so much a part of ourselves that it
seemed impossible that we should be separated.'

In May Thomas himself insisted upon M. Bruch and the English doctor who was also attending giving him a faithful opinion upon his case. They both frankly told him that they feared his disease was incurable. Thomas characteristically pressed for a mathematical statement of the probabilities of his living a year. This the physicians declined to give; but they said openly to him that they deemed the chances to be against his surviving so long. 'Still,' says his sister, 'I do not think he himself took by any means so gloomy a view. Although he knew the possibilities before him, he did not realise the inevitableness or even the probability of the end for some months later, until after November 1884. He would form many schemes for the future; we were to carry out the scheme of the preceding year, and to settle in Australia; or to live at Grasse (near Nice), or in Egypt, after the cholera was over.'

Already, at the Hotel Kirsch, Thomas had been told of wonderful cures of lung disease effected by an English doctor then resident in Paris who claimed to have discovered a new and successful method of treatment. From various sources there came reports of his skill. Thomas made the most careful inquiries and found much to justify faith. The excellent M. Bruch, when consulted, said simply that he could do no more, and that he saw no objection to the trial of a new system of cure. Thomas thereupon entered into correspondence with the physician in question, who insisted upon the necessity of personally seeing his patient. For months there had been debates in the little family as to what place of refuge was to be sought when the arid summer heats began in North Africa with the advent of July. Thomas resolved to go to Paris, and give the much-praised cure a trial. Accordingly, on July 7 the Algerian home was broken up, and a new hegira made northward.

Down to the last the 'slag' experiments were pursued. In June we find Thomas writing to Mr. Gilchrist :—

'I wish I could convince you that our one hope of reducing costs is in slag, as I am sure it is. Remember the phosphorus is more valuable than the iron in pig; only we are too stupid to turn it to account properly.'

On quitting Algiers it was arranged that Mr. Twynam should proceed to Middlesbrough to continue the experiments.

CHAPTER XXII

THE LAST DAYS IN PARIS

CHOLERA and rumours of cholera caused some difficulty in gaining France. ' Quarantine,' says Mrs. Thomas, ' was strict between Algiers and Marseilles. Finding we could go by the Spanish route, we packed hastily and got on board the steamer to Puerto Vendres. The vessel was so full that we could only with much endeavour procure a berth for Sidney. We ourselves were obliged to remain during two nights in the dining saloon, which was so crowded that we could not get even a sofa to ourselves. We all felt very sad at leaving our lovely villa, and parting with so many friends we felt we should never in all probability see again. Through all discomforts on board Sidney was cheerful and hopeful, as he always was in difficulties. We landed early on the morning of July 9. We journeyed by way of Narbonne and Toulouse to Limoges, where we remained a little ; for the intense heat quite exhausted our boy. We arrived in Paris (still gay with the National Fête rejoicings) on July 15 (the morrow of " Bastille Day ") and alighted at the Hôtel Normandie.'

From Limoges Thomas wrote to his old Wiesbaden correspondent :—

To Miss Burton

'Grand Hôtel de la Boule d'Or, Limoges : July 14, '84.

' Dear Bess,—I should have replied before to your kindest of letters ; but the last three or four weeks we have been living in a state of utter uncertainty as to where we should be next week. The cholera scare infected Algiers badly, and finally not only delayed our start by a week, but forced us to go round by Port Vendres,—far the longest route. With considerable regret, we left our pretty home at El Biar last Tuesday, looking quite its prettiest, with flowers, fig-trees, cacti, aloes, oranges, fruit-trees, and vines. We all concluded we should not be likely ever to live in such a pretty place again. The heat for the past week had, however, been pretty considerable (70° to 80° F.); although we now find it is still greater here.

'Our crossing of 30 hours was uncomfortable enough, —tremendously hot; boat much overcrowded, chiefly with Jewish families; sleep out of question. The mother knocked up, but got over it wonderfully.

'We landed at 5 A.M., and went by train to Narbonne. Queer old place, with a staring new quarter. Stopped there 24 hours; then on to Toulouse, where Twynam left us to return to London,—we stopping 48 hours to rest. Animated busy town; back streets and churches old; rest all new. Interesting country all the way from Narbonne.

'We came on here Saturday, and stop till to-morrow, so as not to be in Paris on the Fête Day. We propose going to hotel at first, and then looking for rooms. We may stop only a week or two, or possibly two months, according as I think Dr. —— has or has not anything useful, and as I can get over some business matters connected with France.

'I have been working a little at Algiers on an investigation which may, or may not, lead to a "discovery," but which has anyhow been very instructive (the main thing). It is a kind of offshoot of my old ideas, but in a different direction. I do not expect it will be finished for a year or two; anyhow it has served as an interest to keep me from stagnating, though it has absorbed a good deal of money.

'I doubt if we return to Algiers; though I like it and the people well, the crossing is trying for the mother, and I doubt the climate suiting me.

'The whole town here is disorganised with the Fête. It never went to bed last night and it seems will not to-night.—Yours,

'S. G. T.'

On arrival in Paris, after a few days, comfortable and airy apartments were secured in the Avenue Marceau, and there Thomas spent the last seven months of his life. 'He now only drove out,' says his mother, 'on very fine days. He continued, 'however, to work — continued his investigations. When he was tired with thoughts of business, I would often read to him by the hour together. With us he was always happy, but various letters from England often troubled him much. His brother, Dr. Llewellyn Thomas, wrote urging us to go home, and expressing his belief that our patient would do quite as well in England as in Paris; but Sidney shrank from the notion, indeed told me that business interviews such as would necessarily follow upon a return to London would kill him at once. After that we said no more of the matter.'

Thomas had at once placed himself under the care of the physician whom he had come to Paris to consult, and for a time he apparently derived some benefit from the ' new treatment '; but the improvement did not last and the end was now but too absolutely certain. He did, however, derive much entertainment and, no doubt, some consequent physical benefit, from the conversation of his doctor, who

x

was a much-travelled Ulysses with a great deal to say for himself. The two would engage in hours-long discussions and arguments, which were a real refreshment to the sick man.

The following epistles give some notion of Thomas's pre-occupations and health during this summer and autumn.

To Mr. Chaloner

'61 Avenue Marceau, Paris: August 1, 1884.

' Dear C.,—The above will be our address till the middle of September. Shall be glad to hear from you. Not feeling very bright, or would write.—Yours,

' S. G. T.

' We are close to our old quarters in 1878 ; the Avenue Marceau *used* to be Av. Josephine.'

'61 Avenue Marceau, Paris: October 17, 1884.

' Dear Chaloner,—It is nearly four months since I have heard from you, though I have written you meantime, notwithstanding a very bad attack on the lungs which floored me completely during August and September, and from which I am still hardly quit. I am, however, going in for a special form of treatment which compels my staying in Paris till the cold forces me to bolt, which may be any week. The treatment is I think doing some good, but I hardly know yet. My illness has naturally led to arrears of correspondence all round—particularly as I have had much business to get through meantime.

' What have you been doing all the time ? . . . Have you done anything in the experiment way ? I have got some rather good results after much delay. I presume

you will have no time for experiments, now Birkbeck has
started on a big scale. . . .

'We are having coldish weather, and I am quite tied
to the house. . . . Things in the way of business very
dull. No orders and awful prices. We do better at North
Eastern Steel Co. than our neighbours; but that is our
only comfort. Writing wearies me, so adieu.'

This last letter is in Sidney's own hand; but much of
his correspondence about this time is written by his sister,
sometimes by his mother. The sands were already running low in the glass.

The little family was not left entirely alone in the
Elysian Fields; many friends came from time to time to
see Thomas; most of them, it may be surmised, with a
foreboding that their visit was a farewell one. Among
others who came at this time were Mr. Vacher (one of
Sidney's old chemical teachers of whom we have spoken
above),[1] and his wife. Mr. Vacher, in a letter to Sidney's
sister, thus speaks of his departed friend and of this
visit:—'What I do possess and value exceedingly is the
very definite and vivid impression made on me by his
personality. Of his intellectual side I can hardly speak,
being but a distant admirer of his talent and splendid
achievement. Of his character I should say that its
distinguishing trait was nobility and highmindedness, that
he was by intuition opposed to all that is ignoble and
petty.

'On one occasion he gave me a lesson. . . . I made
use of the word *cads*, and he received it with such quiet
coldness than I at once saw the incongruity of the notions
implied by it with those ideas of fraternity which were
common to us both. . . .

---

[1] *Ante*, p. 36.

'His public spirit was of the highest order. Poverty, prosperity, sickness, death, none of these disturbed the earnest purposefulness of his life. My wife will never forget the impression made on her, (when we saw him a few times in Paris shortly before he died,) by his calm attitude, cheerful patience, and exceedingly sweet expression. She says that, notwithstanding his pale face and wasted frame, the thought of him is always suggestive to her of strength, and in times of weakness she often likes to call up the vision.'

Business friends also would come to see him, 'and it was wonderful,' says his mother, 'to see how, on such occasions, he would gather himself together and be his old erect keen self, but he would afterwards suffer terribly from reaction;' thus justifying his horror of a return to England and concomitant worrying interviews.

In September his brother Llewellyn Thomas (who was after all to die before him) visited him, and Mrs. Thomas remembers a 'happy although sad' time. 'After this visit Sidney's life became entirely that of an invalid. October was upon us, and the weather seldom permitted him to go out for even the shortest drive.' Yet, although thus imprisoned in a sick room, the only change from which (as was becoming more and more apparent) might be to the grave, Thomas did not lose heart. 'He was always full,' says his sister, 'of quaint sayings and jokes which relieved the heaviness of sick-room life. I think no one, coming into the room, would have imagined the anxieties which lay behind our fun and cheerfulness.'

The workers and their lots were ever in his mind. 'One of his favourite subjects of dreaming in the gloaming' (we quote his sister again), 'after we had despatched our letters for the day was the possibility of building a " model lodging house " which should be really

a model. There was, in the very first place, to be a lift—
for goods at least ;—for the poor women coming to the
Thames Police Court had often told him that one of the
strong objections the poor folk had to " model " dwellings
was the drag upon them, often delicate enough as they
were, of carrying every· scuttle of coals or basket of
provisions to the top of the high buildings. There were
to be conversation rooms and reading rooms for men and
women.'

'Slag' was still an engrossing topic of thought, and
Thomas carried on an elaborate correspondence with the
chemists who were working at the question under his
direction.   In November there came news of the success
of the simple German plan of grinding the slag and then
applying it directly to the earth, and Thomas was in communication concerning the good tidings with Mr. Wrightson
of the North Eastern Steel Works, who sent samples for experiment to his brother, Professor Wrightson of Salisbury.
The experiments thereupon conducted by Professor
Wrightson led the way to the adoption of the new
fertiliser in England. 'I may say,' says his sister, 'that
Sidney some years previously had suggested this mode of
treatment to practical farmers ; but he was assured that
the earth would not assimilate the raw slag. He remained
in interesting correspondence with Mr. Wrightson, Professor Wrightson, and Professor Munro up to the last.
At the same time he did not relax his direction of experiments on other processes, and one of the last matters he
was able to take keen pleasure in was a telegram announcing
results obtained by Mr. Tucker.'

In November he writes to Mr. Gilchrist :—

'I shall be thankful to welcome any method of utilising
the slag by treatment or non-treatment.'

Early in this last month it had at length been finally
determined to move south to Grasse. Tickets had been
taken and a *coupé-lit* secured, and Thomas wrote to Mr.
Chaloner by the hand of his sister :—

'Cold and cholera are driving us from Paris. Address
on and after Saturday to Grand Hotel, Grasse, Alpes
Maritimes.'

At the last moment, however, Thomas had a fresh
access of lung trouble, and his doctor advised that a
journey, even with every precaution, would probably be
fatal to him. From Paris he was destined not to move
again.

The clouds were indeed finally closing in upon the too
short sunshine of his life. In this very November Dr.
Llewellyn Thomas, Sidney's elder brother, died suddenly.
A letter came to the little Parisian household one morning
to say that Dr. Thomas was ill; the same afternoon a
telegram announced his death. Sidney's sister immediately
crossed to London. It would not be fitting to attempt to
describe the gloom in which sat Sidney (unable himself to
quit the Avenue Marceau) and his mother (unable to quit
him) under the shadow of this affliction.

From that day it seemed as if Thomas had in truth
entered upon the pathway leading to the end. His sister
noticed upon her return from her sad journey, that Sidney
in sketching, as his wont was, future plans, always left
himself out of account. His great subject of anxiety
now was that the money he left behind him as the reward
of his inventions and the fruit of his toil should be spent,
(mainly spent, after a modest provision had been made for
the mother and sister who were so dear to him,) upon
bettering and making somewhat easier the hard lives of

the toilers who create all wealth. Over and over again he impressed upon his sister the sacred trust he bequeathed to her. Her discretion as to ways and times—subject to certain general lines which he laid down—was to be absolute; but to the workers the money was in the bulk to go. His mother he would urge, as she says, 'almost passionately,' to husband her remaining vitality, that she might live to help and strengthen his sister in her task.

If ever there were a logical consistent life that life was Thomas's. The old boyish dream of making a fortune had been realised as few dreams are, and (a far more wonderful thing) the old boyishly imagined use to which that fortune was to be put, the aid and comfort of the needy and the oppressed, was to be realised too. A nature uncorrupted by the 'deceitfulness of riches' is a rare one indeed.

It is right to say here that Thomas, unlike some of us who, for weal or for woe, have become distrustful of old faiths, was a firm believer in immortality. 'He was perfectly persuaded,' says his sister, 'of a future existence. During these last days he would say to Mother, "You I shall see soon, dear Mother; but you, Lily, not for some fifty years yet." He held, too, quite as firmly, that he should be conscious of what we were doing here while he was waiting for us. In the dusk of the evening he would speculate, as we sat together, on the possibility of his manifesting himself to us whom he so dearly loved.'

The new year of 1885 opened sadly enough for those in the Avenue Marceau, who now knew but too well that they could only wait for the end. 'Sidney,' says his mother, 'only grew, the nearer that end approached, more gentle, patient, and thoughtful, and more anxious to ease the parting to us.' In these last days his devoted nurses would get Thomas up as of old and settle him in his chair with his books and papers by his side, but it was little he could write, and that laboriously. Most of his books even were too heavy for him to hold. His sister would read much to him.

In the middle of January his surviving brother Arthur, now a fully qualified medical man, was summoned from his professional duties to Paris, and his skilled and brotherly care and help were greatly prized by the dying man.

The 'end' came at length. In the early morning of February 1, 1885, Sidney Thomas died quietly in his sleep, in the presence of his mother, sister, and brother, breathing only two or three heavy sighs. The immediate cause of death was emphysema.

He had, by his great invention, left a far more permanent mark upon the world than many a veteran general or aged 'statesman;' but he had not completed his thirty-fifth year. He was buried, by his own earnest desire, in the Passy Cemetery. He had shown much distaste to the notion of his dear ones crossing the Channel with his body in the dreary winter weather.

His mother concludes the notes for her son's life from which we have so often quoted by a citation from Jeremy Taylor :—

'It is a vast work any man may do, if he never be idle; and it is a huge way a man may go in virtue, if he never goes out of his way by a vicious habit or a great crime. Strive not to forget your time, and suffer none of it to pass undiscerned. So God dresses us for Heaven.'

313

CONCLUSION

THE old hackneyed, yet ever new and ever untranslatable, Virgilian line,—fraught with all that 'tender majesty' which makes the Latin singer dearer to us than even greater poets,—rises perforce to our memory as we contemplate the death of this young inventor at an age when many have scarcely entered upon their life-work, and as we dream of so much he might have done in the world, perchance upon quite different lines—

Sunt lacrymæ rerum et mentem mortalia tangunt.

No moral needs to be tagged to a memoir of Sidney Thomas. His is a life which speaks for itself.

It only remains to add that the great process of steel manufacture with which his name will be ever identified has thriven and flourished as he expected it to thrive and flourish. As we have seen, in 1878 there was not even in existence any public record of successful dephosphorisation of pig iron. In 1884, 864,000 tons of basic steel were produced. In 1890 the production was 2,603,083 tons. Moreover, in this last year, too, there were also produced, together with the steel, 623,000 tons of slag, most of which was used for fertilising purposes.

Thomas's plans for the disposal of his money for the benefit of the toilers have (it is probably unnecessary to

Y

say), been carried out by those loved ones whom he left behind, and many a life has been gladdened by the results of his labours. Truly his short life has a completeness lacking to many long ones, and of him it may indeed be said : *Finis coronat opus.*